Selected Titles in This Series

682 **Martin Majewski,** Rational homotopical models and uniqueness, 2000

681 **David P. Blecher, Paul S. Muhly, and Vern I. Paulsen,** Categories of operator modules (Morita equivalence and projective modules), 2000

680 **Joachim Zacharias,** Continuous tensor products and Arveson's spectral C^*-algebras, 2000

679 **Y. A. Abramovich and A. K. Kitover,** Inverses of disjointness preserving operators, 2000

678 **Wilhelm Stannat,** The theory of generalized Dirichlet forms and its applications in analysis and stochastics, 1999

677 **Volodymyr V. Lyubashenko,** Squared Hopf algebras, 1999

676 **S. Strelitz,** Asymptotics for solutions of linear differential equations having turning points with applications, 1999

675 **Michael B. Marcus and Jay Rosen,** Renormalized self-intersection local times and Wick power chaos processes, 1999

674 **R. Lawther and D. M. Testerman,** A_1 subgroups of exceptional algebraic groups, 1999

673 **John Lott,** Diffeomorphisms and noncommutative analytic torsion, 1999

672 **Yael Karshon,** Periodic Hamiltonian flows on four dimensional manifolds, 1999

671 **Andrzej Rosłanowski and Saharon Shelah,** Norms on possibilities I: Forcing with trees and creatures, 1999

670 **Steve Jackson,** A computation of δ_5^1, 1999

669 **Seán Keel and James M^c^Kernan,** Rational curves on quasi-projective surfaces, 1999

668 **E. N. Dancer and P. Poláčik,** Realization of vector fields and dynamics of spatially homogeneous parabolic equations, 1999

667 **Ethan Akin,** Simplicial dynamical systems, 1999

666 **Mark Hovey and Neil P. Strickland,** Morava K-theories and localisation, 1999

665 **George Lawrence Ashline,** The defect relation of meromorphic maps on parabolic manifolds, 1999

664 **Xia Chen,** Limit theorems for functionals of ergodic Markov chains with general state space, 1999

663 **Ola Bratteli and Palle E. T. Jorgensen,** Iterated function systems and permutation representation of the Cuntz algebra, 1999

662 **B. H. Bowditch,** Treelike structures arising from continua and convergence groups, 1999

661 **J. P. C. Greenlees,** Rational S^1-equivariant stable homotopy theory, 1999

660 **Dale E. Alspach,** Tensor products and independent sums of $\mathcal{L}_p$-spaces, $1 < p < \infty$, 1999

659 **R. D. Nussbaum and S. M. Verduyn Lunel,** Generalizations of the Perron-Frobenius theorem for nonlinear maps, 1999

658 **Hasna Riahi,** Study of the critical points at infinity arising from the failure of the Palais-Smale condition for n-body type problems, 1999

657 **Richard F. Bass and Krzysztof Burdzy,** Cutting Brownian paths, 1999

656 **W. G. Bade, H. G. Dales, and Z. A. Lykova,** Algebraic and strong splittings of extensions of Banach algebras, 1999

655 **Yuval Z. Flicker,** Matching of orbital integrals on $GL(4)$ and $GSp(2)$, 1999

654 **Wancheng Sheng and Tong Zhang,** The Riemann problem for the transportation equations in gas dynamics, 1999

653 **L. C. Evans and W. Gangbo,** Differential equations methods for the Monge-Kantorovich mass transfer problem, 1999

652 **Arne Meurman and Mirko Primc,** Annihilating fields of standard modules of $\mathfrak{sl}(2,\mathbb{C})^\sim$ and combinatorial identities, 1999

(*Continued in the back of this publication*)

Rational Homotopical Models and Uniqueness

MEMOIRS
of the
American Mathematical Society

Number 682

Rational Homotopical Models and Uniqueness

Martin Majewski

January 2000 • Volume 143 • Number 682 (end of volume) • ISSN 0065-9266

American Mathematical Society
Providence, Rhode Island

1991 *Mathematics Subject Classification.*
Primary 55P62; Secondary 55U35, 57T05, 57T30, 17B70, 20F18, 18D10, 18G30, 55U10, 55U15.

Library of Congress Cataloging-in-Publication Data

Majewski, Martin 1963–
Rational homotopical models and uniqueness / Martin Majewski.
p. cm. — (Memoirs of the American Mathematical Society, ISSN 0065-9266 ; no. 682)
"January 2000, volume 143, number 682 (end of volume)."
Includes bibliographical references.
ISBN 0-8218-1920-8 (alk. paper)
1. Homotopy theory. 2. Hopf algebras. I. Title. II. Series.
QA3 .A57 no. 682
[QA612.7]
510 s–dc21
[514′.24] 99-049715

Memoirs of the American Mathematical Society

This journal is devoted entirely to research in pure and applied mathematics.

Subscription information. The 2000 subscription begins with volume 143 and consists of six mailings, each containing one or more numbers. Subscription prices for 2000 are $466 list, $419 institutional member. A late charge of 10% of the subscription price will be imposed on orders received from nonmembers after January 1 of the subscription year. Subscribers outside the United States and India must pay a postage surcharge of $30; subscribers in India must pay a postage surcharge of $43. Expedited delivery to destinations in North America $35; elsewhere $130. Each number may be ordered separately; *please specify number* when ordering an individual number. For prices and titles of recently released numbers, see the New Publications sections of the *Notices of the American Mathematical Society.*

Back number information. For back issues see the *AMS Catalog of Publications.*

Subscriptions and orders should be addressed to the American Mathematical Society, P. O. Box 5904, Boston, MA 02206-5904. *All orders must be accompanied by payment.* Other correspondence should be addressed to Box 6248, Providence, RI 02940-6248.

Memoirs of the American Mathematical Society is published bimonthly (each volume consisting usually of more than one number) by the American Mathematical Society at 201 Charles Street, Providence, RI 02904-2294. Periodicals postage paid at Providence, RI. Postmaster: Send address changes to Memoirs, American Mathematical Society, P. O. Box 6248, Providence, RI 02940-6248.

This publication is indexed in *Science Citation Index*®, *SciSearch*®, *Research Alert*®, *CompuMath Citation Index*®, *Current Contents*®/*Physical, Chemical & Earth Sciences.*
Printed in the United States of America.

∞ The paper used in this book is acid-free and falls within the guidelines established to ensure permanence and durability.
Visit the AMS home page at URL: `http://www.ams.org/`

10 9 8 7 6 5 4 3 2 1 05 04 03 02 01 00

Table of Contents

Abstract *x*
Keywords *x*
Preface *xi*
Introduction *xiii*

1. Homotopy Theory **1**
1. Homotopical Categories *1*
  1. The axioms *1*
  2. Left homotopical categories *3*
  3. Homotopical subcategories *4*
2. Fundamental Results *6*
  1. Lifting and extension *6*
  2. The derived category *7*
  3. Homotopical functors and their derived functors *8*
  4. The Adjoint Functor Theorem *9*
3. Comonoids up to Homotopy *11*
  1. ... as comonoids over the derived category *11*
  2. Derived tensor product *12*
  3. Generalizations *13*
A. Examples of Homotopical Categories *15*
  1. Cofibration categories *15*
  2. Model categories *18*
  3. Spaces *20*
  4. Simplicial objects *21*

2. Differential Algebra **25**
1. Preliminaries *25*
  1. Chain complexes *25*
  2. DG (co)algebras *27*
  3. Tensor (co)algebras *29*
2. Twisting Maps and the (Co)Bar Construction *30*
  1. Twisting maps and homotopies *30*
  2. The (co)bar construction *30*
  3. Compatibility with tensor product *32*
  4. Homological properties *32*
3. Acyclic Models *33*
  1. Representable functors *33*
  2. The method of acyclic models *35*
  3. Duality *36*
  4. Acyclic model theorems for twisting maps *40*
4. EZ-Morphisms *43*
  1. Extension of an EZ-morphism *43*
  2. A generalization *44*
  3. Properties of the extension *46*

B. CHAIN (CO)FUNCTORS *48*
1. Monoidal categories *48*
2. Normalization *49*
3. Representable cofunctors for spaces *51*
4. Cohomology theories *55*

3. Complete Algebra **57**
1. COMPLETE AUGMENTED ALGEBRAS *57*
1. Ring systems *57*
2. Complete modules *59*
3. Complete augmented algebras and free groups *63*
4. Rigidity *65*
2. COMPLETE LIE ALGEBRAS AND COMPLETE HOPF ALGEBRAS *67*
1. Complete Hopf algebras and the exponential mapping *67*
2. The PBW–Theorem *69*
3. Normal complete Hopf algebras *73*
4. Rigidity *76*
3. COMPLETE GROUPS *77*
1. Nilpotent groups *77*
2. Complete groups *79*
3. The Lazard - Mal'cev correspondence *82*
4. The Quillen functor *86*
C. FILTERED MODULES *87*
1. Filtered vs. cofiltered modules *87*
2. Normal maps and exactness *89*
3. Filtered tensor product *91*
4. Complete Differential Algebra *94*

4. Three Models for Spaces **97**
1. THE CELLULAR MODEL *97*
1. The homotopical category of dg algebras *97*
2. The homotopical category of dg Hopf algebras up to homotopy *99*
3. The cobar - chain functor and the chain - loop functor *100*
4. Compatibility with (tensor) products *101*
5. The homotopy diagonals *102*
2. THE SULLIVAN MODEL *104*
1. The homotopical category of commutative dg* algebras *104*
2. The Sullivan cofunctor and Stokes' map *106*
3. Extension of Stokes' map *108*
4. Compatibility with (tensor) products *109*
5. Dualization *110*
6. The homotopy diagonals *113*
3. THE QUILLEN MODEL *115*
1. The homotopical category of dg Lie algebras *115*
2. The Quillen functor *116*
3. Connection to the chain - loop functor *118*
4. The group algebra of a free simplicial group *121*
5. A proof of the Quillen equivalence *123*

4. MAIN RESULTS *126*
1. Summary *126*
2. Anick's equivalence *127*
3. Proof of the Baues-Lemaire conjecture *128*
4. Rational equivalence *129*
D. THE CELLULAR LIE ALGEBRA MODEL *130*
1. A natural diagonal for the cobar-chain functor *130*
2. A natural Hopf diagonal *133*
3. The category of dg Lie algebras (over any ring) *134*
4. Anick's theorems and naturality *139*
NOTATIONS *145*
BIBLIOGRAPHY *147*

Abstract
The main goal of this paper is to prove the following conjecture of Baues and Lemaire: the differential graded Lie algebra associated with the Sullivan model of a space is homotopy equivalent to its Quillen model. In addition we show the same for the cellular Lie algebra model which we build from the simplicial analog of the classical Adams - Hilton model. It turns out that this cellular Lie algebra model is one link in a chain of models connecting the models of Quillen and Sullivan. The key result which makes all this possible is Anick's correspondence between differential graded Lie algebras and Hopf algebras up to homotopy. In addition we show that the Quillen model is a rational homotopical equivalence, and we conclude the same for the other models using our main result. The construction of the three models is given in detail. The background from homotopy theory, differential algebra, and algebra is presented in great generality.

Keywords
Rational homotopy theory, Quillen model, Sullivan model, Adams and Hilton model, Anick equivalence, Hopf algebra up to homotopy, Baues and Lemaire conjecture, abstract homotopy theory, model category, cofibration category, differential homological algebra, tensored category, monoidal category, filtered module, complete module, ring system, Lie algebra, Hopf algebra, nilpotent group, completion, Malcev correspondence, SDR-data, EZ-morphism, cobar construction.

Foreword

The work on this paper started with the observation that the Baues - Lemaire conjecture (1978) could be proved with the help of Anick's beautiful and unexpected results on Hopf algebras up to homotopy (1988). Since it was formulated, there were many announcements of proof of the conjecture, but as far as I know there still is no proof available. Also recent attempts to find a simpler proof than the one given here have been unsuccessful.

It is now some years ago that I first published a sketch of my proof of the conjecture ([Maj 2]). Since then, I spent a lot of time in trying to order the pieces. In doing so I included more and more background material, so that now the text is nearly self-contained. The background is presented in great generality so as to be applicable also in other contexts.

I hope that the reader will profit from the systematic (mostly categorical) presentation. Since we recall here many auxiliary results and proofs which in an almost sufficient form are spread over the literature, the present text is of course far from being a *minimal* presentation of the subject. The reader who does not want to delve into three different theories before coming to the point, I suggest to begin with the introduction and proceed with Chapter 4.

The manuscript of the present text was submitted to the Freie Universität Berlin in February 1996 as doctoral thesis.

Introduction

From a very general point of view, the subject of this work is analogous to a quite common scheme:

Suppose given two translations T_1 and T_2 of some basis language into say language one and language two. Futher let there be given a translation U from language one into language two. Now there are obviously two translations of the basis language into language two: T_2 and $U \circ T_1$ (the second denoting a composition of two translations). Does these translations give the same, at least up to some notion of similarity? Of course, if we require translations to be "meaning-preserving", the answer is yes. However, it is hard to define meaning and (in the case of natural languages) even harder to preserve it.

In the mathematical context, the languages are categories, the translations are functors, and the similarity is a homotopical structure on the category. More specifically, the basis category is that of spaces and we consider the two well-known rational models for spaces. Does they give the same via the obvious functor U (up to weak equivalence)? It is the motivation and main purpose of this work to show that this is indeed the case.[1]

Rational homotopy theory

The subject of rational homotopy theory was established by Quillen in his pioneer work from 1969. Using ideas of classical Lie theory, he introduced a certain functorial construction associating with any 1-connected topological space a dg Lie algebra over the rationals. He proved that this model reflects faithfully the rational homotopy type of the space. Some years later Sullivan constructed another model and proved a similar result. He used a variant (due to Thom) of the complex of differential forms on a manifold. The model is thus a commutative dg* algebra over the rationals. Both the Quillen and the Sullivan model were used to solve geometric problems by algebraic machinery.

There is another well-known and much older (1955) model for spaces, the Adams-Hilton model, which is a dg algebra. In 1988 Anick extended the structure of this model by taking into account any (multiplicative) diagonal approximation. The result is what he called a (dg) Hopf algebra up to homotopy. He succeded to show that from this model one can construct a dg Lie algebra model.

The Baues-Lemaire conjecture

The Sullivan model (of a given space), being a commutative dg* algebra, has an associated dg Lie algebra. So it is natural to ask whether this dg Lie algebra model is homotopy equivalent to Quillen's.

Our main goal is to show that this is indeed the case. This proves in the affirmative a conjecture of Baues-Lemaire from 1978. It was known that the two models have many properties in common, and by computation the conjecture was proved for

[1]Received by the editor July 3, 1996.

rationally very simple spaces, e.g. spheres and Eilenberg - Mac Lane spaces, where one does not have to go into the actual construction of the models.[2]

It was also conjectured, by Anick and others, that his "Adams - Hilton - Anick" model is equivalent to Quillen's. It turns out that a simplicial variant of Anick's model, which we introduce here as the *cellular model*, lies "between" Quillen's and Sullivan's. So it appeared natural to try to show that actually all the three models are homotopy equivalent. This is what we are going to do.

Method of proof

We use Anick's key theorem: the universal enveloping functor is a homotopical equivalence between the categories of dg Lie algebras and dg Hopf algebras up to homotopy. We use this equivalence to reduce the initial problem: In order to prove certain dg Lie algebras equivalent, it suffices to do the same for their universal enveloping algebras — regarded as dg Hopf algebras *up to homotopy* (and not as dg Hopf algebras, which is equivalent to the original problem).

For more details see the summary of Chp. 4 below.

Organization

The main results of this paper are all presented in the last Chp. 4, while the first three chapters constitute the background in form of three (beginnings of) stand-alone theories, containing the technical results needed in the proof. Each chapter has an appendix. Moreover at the end of the text there is a list of notations resp. abbreviations concerning categories. Other conventions resp. terminologies are explained later in this introduction and at the beginning of each chapter.

We next give a summary of each of the four chapters.

Chapter 1

We present a generalization of the well-known axiomatic frameworks for *Homotopy Theory* (due to Quillen resp. Baues) by introducing *homotopical categories*. This is motivated by the following question.

Fibrations and limits (resp. the dual notions) often play a minor role in certain applications of axiomatic homotopy theory, e.g. in this work. Moreover in some contexts this sort of structure does not exist, e.g. when dealing with objects endowed with some up-to-homotopy structure. — Is it possible to create an axiomatic framework without using fibrations and limits, and which applies to objects with up-to-homotopy structure?

We show that such objects can be treated conveniently within the framework of homotopical categories. More precisely, we discuss the case of *comonoids up to homotopy* which specializes to dg Hopf algebras up to homotopy. Another useful new feature is the possibility of a reasonable definition of homotopical *sub*categories. The two first sections may also be read as a summary of the relevant (to us) aspects of Quillen's model categories resp. Baues' cofibration categories. These are recalled in App. A, and the new point of view leads to some new results. Here we also discuss the homotopical category of simplicial objects over some category, based on Quillen's work on the same subject.

[2] It is interesting to note that a negative answer to the conjecture would yield a non-trivial automorphism of the rational homotopy category of 1-connected spaces (of finite rational type). Of course it remains open whether such a curious functor exist.

Chapter 2

We give an introduction to *Differential Algebra*. Our treatment is purely categorical, we work over any additive monoidal category (playing the role of K-modules). In this context we present most of the various techniques and auxiliary results needed in the "concrete" part of the text. One finds sign conventions and notations, the cobar construction and twisting maps (= "twisting cochains"), the method of acyclic models (with some new features) and EZ-morphisms (= "SDR-data"). The results are in any case generalizations and in some cases extensions of more or less known results.

In the appendix we give the categorical background (monoidal categories, functors and transformations) and recall the fundamental relationship between simplicial sets and differential algebra. This culminates in a surprisingly general and far reaching treatment of cohomology theories (the Sullivan cofunctor is an example). The main result shows that the "theories cohomologiques" (in the sense of Cartan) are precisely the cofunctors from simplicial sets to dg^* algebras which are representable and acyclic on models.

It seems to me that the reading is not complicated by the categorical setting. One may think of modules with their usual tensor product, but gets much more. In contrary, the perspective often gives clear insight. For instance, we treat *categorical duality* (opposite category) and *functional duality* (Hom-cofunctors) simultaneously in terms of adjoint cofunctors.

We almost completely ignore in this context Lie algebras and commutative algebras. This is mainly because the most interesting results must depend on certain divisibility conditions. In Chp. 3, App. C, N°4 we suggest a categorical treatment of the (anti-) commutative theory.

Chapter 3

The last discipline of the general kind is what we call *Complete Algebra*. In abstract terms, this is the study of various species of algebras — over the additive monoidal category of complete modules (i.e. algebraic- and topologically complete decreasing filtered modules). Here we do not explicitly work in the DG setting. However, unless complete groups are involved, everything is easily generalized to the more general context (we have chosen the arguments accordingly).

We extend Quillen's work [Q | App. A] in some respect. In particular, we replace the ground field of characteristic 0 by any mild ring system, i.e. a sequence of rings obtained from any (commutative, as always) ring by successively inverting primes, at least those $\leqslant n$ after n steps. We also omit the condition that "$\mathrm{gr}(A)$ is generated by $\mathrm{gr}^1(A)$" which roughly means that the filtration of A is closest to the minimal filtration as algebra resp. group (augmental filtration resp. lower central series). This *rigidity* assumption is annoying in the DG context; omitting it we are forced to differ significantly from Quillen's exposition.

The key result is a version of the PBW-Theorem which we prove in full generality. Using it we give a new proof of the Lazard correspondence between complete groups and complete Lie algebras, extended by a characterization of those complete Hopf algebras that correspond precisely to complete Lie algebras, hence to complete groups (over a field these are *all* complete Hopf algebras).

Some of the material is not applied elsewhere in this paper, in particular most of § 3 and App. C. The general setting (use of ring systems) and the systematic study of filtered modules and their tensor product (within the DG setting) will be of importance in the extension of our main results to mild homotopy theory, see "Extensions" below.

Chapter 4

We discuss in some detail the *Three Models for Spaces* which we want to compare and the corresponding homotopical - algebraic theories:

(i) the cellular model as a dg algebra resp. Hopf algebra up to homotopy,
(ii) the Sullivan model as a commutative dg* algebra and
(iii) the Quillen model as a dg Lie algebra.

Of course we concentrate on questions related to the connection between these models. The main results of the first three sections are Theorem–A (§ 1), Theorem–B (§ 2) and Theorem–C (§ 3). These results are reformulated in § 4, where we show that by employing Anick's equivalence the desired Baues - Lemaire conjecture easily follows.

The conjecture states the following: Let the coefficient ring K be the field of rationals.[3] Let X be a 1-connected space[4] of finite rational type (i.e. with finite Betti-numbers). Let A^*X be the Sullivan - Thom algebra, i.e the commutative dg* algebra of polynomial forms on X. Choose an H^*-equivalence[5] $M \xrightarrow{\sim} A^*X$ with M minimal (or at least 2-reduced of finite type). Consider the dg Lie algebra $\mathcal{L}(M) = \mathcal{P}\Omega(\#M)$.[6] On the other hand let $L \xrightarrow{\sim} \lambda GX$ be a minimal model of Quillen's dg Lie algebra construction on X. Does the two described constructions yield the same result, at least up to weak equivalence?

BAUES - LEMAIRE CONJECTURE. *There is a weak equivalence between the dg Lie algebras $\mathcal{L}(M)$ and L. Hence any minimal model of $\mathcal{L}M$ is isomorphic with L.*

Equivalently, by Milnor - Moore, one may try to construct a weak equivalence of dg Hopf algebras between $\Omega(\#M)$ and UL. That is, we need a weak equivalence of dg algebras which is also compatible with the obvious (coalgebra) diagonals.

The first part of this (the equivalence as dg algebras) is already non-trivial. Essentially, we prove this by constructing a (finite) sequence of dg algebras and of weak equivalences from one to the next, or conversely (see diagram below). One of these intermediate dg algebras is $\Omega C(X)$, which we call the cellular model. It is sort of a (natural) Adams - Hilton model for X. Thus, without having planned this from the beginning, we are faced with the problem of comparing all the three "classical" models for spaces.

The second part of the problem (compatibility with diagonals) cannot be solved this way. In fact, some dg algebras in the chain does not seem to have a significant Hopf diagonal. And even if they had, it is not expectable that the maps, being canonical only up to homotopy, commute strictly with them. In this misery, Anick's work brings us a big step forward: the compatibility with diagonals follows from the *compatibility up to derivation homotopy* (if the underlying objects are free).

[3]It suffices that K is any hereditary $\mathbb{Q}$-algebra. Recall that a ring is *hereditary* if the submodule of any free module is free. The commutative hereditary rings are precisely the P.I.D.s. Of course, we only consider commutative coefficient rings.

[4]We use "space" for simplicial set. One may think of the singular simplicial complex of a topological space.

[5]An F-equivalence is by definition a morphism inducing an isomorphism via the functor F.

[6]Here Ω is the cobar construction and $\#$ the Hom-dualization cofunctor.

This suggests a new strategy: construct on each of the dg algebras connecting $\Omega(\#M)$ and UL a *homotopy diagonal* and prove that each map in the chain above is compatible with this structure (up to homotopy). The point is that a homotopy diagonal need not be map $A \to A \otimes A$ in $\langle \mathrm{dgA} \rangle_0$, we only require a morphism in the derived category $\mathrm{Ho}\langle \mathrm{dgA} \rangle_0$. Here the general framework of homotopical categories is useful. In fact, the described procedure leads to the "correct" general notion of dg Hopf algebra up to homotopy, in a context where the (underlying) algebras need not be free (§ 1, N° 2).

$\Omega(\#M)$	loop algebra of the Sullivan model	
$\uparrow \sim$		
$F^*A^*(X)$	"loop algebra" of the Sullivan - Thom algebra	
$\uparrow \sim$		THEOREM–B (4.51)
$\Omega C(X)$	cellular algebra model (cobar - chain functor)	
$\downarrow \sim$		THEOREM–A (4.18)
$C(GX)$	chain algebra of the loop space (chain - loop functor)	
$\downarrow \sim$	}	
$N(\widehat{\mathsf{K}}_{.}GX)$	}	THEOREM–C (4.61)
$\uparrow \sim$	}	
$U\lambda(GX)$	universal enveloping algebra of Quillen's Lie algebra	
$\uparrow \sim$		
UL	universal enveloping algebra of the Quillen model	

The chain of weak equivalences.

Theorems–A, B, C provide a chain of weak equivalences between the following constructions (viewed as models) for a space X. Except for the first and the last model, which are described above, these are functorial. The arrows are morphisms of dg Hopf algebras up to homotopy (dgHh-maps), i.e. they are morphisms of dg algebras (dgA-maps) which commute, in the derived category, with the canonically defined homotopy diagonals.

The loop algebra construction F^* is essentially the Hom-dual of the Bar construction. In case of M it coincides with $\Omega\#$. In order to show that the second arrow is a weak equivalence we have to assume that the space X is of finite rational type. This is not surprising, because of the implicit double-dualization in $F^*A^*(X)$.

The (homotopy) diagonal on the cellular algebra model $\Omega C(X)$ is not obvious, we consider it more closely in App. D, N° 1. On $C(GX)$ we consider the Alexander - Whitney diagonal (which fortunately *is* a dgA-map).

The Quillen's Lie algebra functor can be written as a composite $N\,\mathcal{P}_{.}\widehat{\mathsf{K}}_{.}G$; we abbreviate the composite of the first three functors by λ. Here $\widehat{\mathsf{K}}$ produces the completion (with respect to the augmental filtration) of the group ring of a group and $\mathcal{P}$ assigns the primitive Lie algebra to a complete Hopf algebra. The subscript points indicate that these functors are applied degreewise to simplicial objects.

Finally N is the usual normalization, which carries simplicial Lie algebras to dg Lie algebras.

Corollaries.

As sort of a supplement we construct a (significant) *natural* Hopf diagonal on the cobar construction of the chain coalgebra of a 2-reduced space, i.e. the cellular algebra model (App. D). This yields a (natural!) *cellular Lie algebra model* for spaces, where cellular means that the generators in that free dg Lie algebra correspond to the cells (non-degenerate, of strictly positive dimension) of the space. It is even possible, at least in principle, to compute the differential by means of the attaching maps of the cells. Here the Hausdorff formula comes in.

We also give (in § 3, N° 5) a proof of the Quillen (rational homotopical) equivalence. Our proof is somewhat more elementary, in that we use our weak equivalence $C(GX) \to N(\widehat{\mathsf{K}}.GX)$ where Quillen employs Curtis' hard connectivity results for the terms of the lower central series of a free simplicial group (the so-called convergence theorem). By virtue of our main result the Quillen equivalence implies that also the cellular model (and the Sullivan model) constitute rational homotopical equivalences (§ 4, N° 4).

Extensions

There exist refinements of the Quillen model and the Sullivan model where the ground ring (the rationals) is replaced by a sufficiently fast increasing ring sytem. Let me indicate this as follows: Quillen ⇒ Dwyer (1972) [Dw] [ShT2], Sullivan ⇒ Cenkl - Porter (1980) [CP] ⇒ FU Berlin (1986) [FU] ⇒ Majewski - Stelzer (1988) [Maj] [Ste] ⇒ Scheerer - Schuch - Vogt (1991) [FU2]. There is nothing to improve on the cellular model, for it is already defined over the integers. Using the work of Chp. 1, Chp. 2 and Chp. 3, I have proved analogs of Theorems–A, B, C for the finer models, i.e. in the context of mild homotopy theory.[7] A suitable generalization of Anick's equivalence[8] would thus imply a generalization of our main result, proving some *mild Baues - Lemaire conjecture*. I leave it's verification to others.

[7] In fact, the generality of these chapters is mainly a consequence of these efforts!

[8] In [Hess] such a generalization is stated, but it seems insufficient.

Chapter 1

Homotopy Theory

Let $\mathcal{C}$ be any category. Then $X \in \mathcal{C}$, or $X \in \mathrm{Ob}\,\mathcal{C}$, means X is an object, while $f \in \mathrm{mor}\,\mathcal{C}$, or $f : X \to Y \in \mathcal{C}$, means f is a map (= morphism) in $\mathcal{C}$ from X to Y. Given two such maps $f, g : X \to Y$ we also write $f, g : X \rightrightarrows Y$. A contravariant functor is also called a cofunctor; a natural transformation is also called a natural map. We use lim to denote the limit of a diagram, and colim to denote the colimit.

§ 1 Homotopical Categories

N° 1 The axioms

Let $\mathcal{C}$ be a category and $we \subset \mathrm{mor}\,\mathcal{C}$ a distinguished class of maps in $\mathcal{C}$ which are called *weak equivalences*. We use the notation $f : X \xrightarrow{\sim} Y$ to indicate that f is a weak equivalence.

Let $\mathcal{C}_c$, $\mathcal{C}_f \subset \mathrm{Ob}\,\mathcal{C}$ be classes of objects. An object $X \in \mathcal{C}$ is said to be *cofibrant* resp. *fibrant* resp. *cofibrant - fibrant*, if X lies in $\mathcal{C}_c$ resp. $\mathcal{C}_f$ resp. $\mathcal{C}_{cf} := \mathcal{C}_c \cap \mathcal{C}_f$. We also regard $\mathcal{C}_c$, $\mathcal{C}_f$ and $\mathcal{C}_{cf}$ as full subcategories of $\mathcal{C}$.

Definition (1.1)

We call $(\mathcal{C}, we, \mathcal{C}_c, \mathcal{C}_f)$ a *homotopy structure* in $\mathcal{C}$ if the axioms (H-1), (H-2), (H-3) and (H-4) are satisfied, see below. We also call $\mathcal{C}$, endowed with a distinguished homotopy structure, a *homotopical category*.

Axiom (H-1) requires that weak equivalences have the usual "2 $\Rightarrow$ 3"-property. Axiom (H-2) requires the existence of cofibrant and of fibrant models. By axiom (H-3) there is a so-called 'derived' category, which turns out to be the localization of $\mathcal{C}, we$. Axiom (H-4) requires the existence of cylinder objects and path objects. In the present chapter, the last axiom (H-4) plays an essential role only in connection with homotopical subcategories, see N° 3. In particular it is not needed to obtain the fundamental results of § 2. However cylinder objects resp. path objects are important in practice.

(H-1) If f, g are maps in $\mathcal{C}$ such that gf is defined, then if two of f, g and gf are weak equivalences, so is the third.
Isomorphisms are weak equivalences.

(H-2) Each object X has a *cofibrant model* $\ell : LX \xrightarrow{\sim} X$ (i.e. LX is cofibrant and $\ell \in we$) and a *fibrant model* $r : X \xrightarrow{\sim} RX$ (i.e. RX is fibrant and $r \in we$), such that:
X fibrant $\Rightarrow LX$ fibrant, X cofibrant $\Rightarrow RX$ cofibrant.

Before stating the next axiom we introduce two relations for certain maps in $\mathcal{C}$. Let X be a cofibrant object. We define a *left homotopy object* for X to be a cofibrant object $\tilde{X}$ together with maps $i_0, i_1 : X \rightrightarrows \tilde{X}$, $q : \tilde{X} \to X \in \mathcal{C}$ such that $qi_0 = \mathrm{id} = qi_1$ and q is a weak equivalence; then by (H-1) also i_0, i_1 are weak equivalences.

Now let $f, g : X \rightrightarrows Y \in \mathcal{C}$. We call f *left homotopic to* g ($f \overset{\ell}{\sim} g$), if there is a left homotopy object $(\tilde{X}, i_0, i_1, q)$ and a map $H : \tilde{X} \to Y$ with $Hi_0 = f$, $Hi_1 = g$. Such H is called a *left homotopy* from f to g.

By duality there is a *right homotopy* relation ($f \overset{r}{\sim} g$) based on the notion of *right homotopy object* $(\tilde{Y}, p_0, p_1, s)$ for Y (here we assume Y to be fibrant, while X is arbitrary).

These relations are clearly reflexive and symmetric, but not transitive in general.

(H-3) There is a category $\mathrm{Ho}\,\mathcal{C}$ and a functor $\gamma : \mathcal{C} \to \mathrm{Ho}\,\mathcal{C}$ with the following properties:

(o) γ is the identity on objects and carries weak equivalences to isomorphisms.

(i) Let B be cofibrant and Y fibrant. Then for every morphism $\varphi : B \to Y \in \mathrm{Ho}\,\mathcal{C}$ there is some $f : B \to Y \in \mathcal{C}$ with $\gamma(f) = \varphi$.

(ii) Let B, Y be as above. Then for all $f, g : B \rightrightarrows Y$ we have:

$$f \overset{\ell}{\sim} g \Leftrightarrow \gamma(f) = \gamma(g) \Leftrightarrow f \overset{r}{\sim} g.$$

These three are the most important axioms. In § 2, N° 2 it is shown that by virtue of these axioms $\gamma : \mathcal{C} \to \mathrm{Ho}\,\mathcal{C}$ is a localization of $\mathcal{C}$ with respect to the class of weak equivalences. In particular, $\mathrm{Ho}\,\mathcal{C}$ is unique up to canonical isomorphism of categories. Notice that the first property (H-3) already implies that γ sends left (resp. right) homotopic maps in $\mathcal{C}$ to the same map in $\mathrm{Ho}\,\mathcal{C}$.

Definitions (1.2)

Any (fixed) $\gamma : \mathcal{C} \to \mathrm{Ho}\,\mathcal{C}$ as in (H-3) is called the *derived category* of $\mathcal{C}$. It is customary (though unprecise) to write $\mathrm{Ho}\,\mathcal{C}$ for the derived category.

A diagram in $\mathcal{C}$ is *commutative up to homotopy*, or *commutes in* $\mathrm{Ho}\,\mathcal{C}$, if its image under γ is a commutative diagram in $\mathrm{Ho}\,\mathcal{C}$.

We use the notation $[A, X] = \mathrm{Ho}\,\mathcal{C}(A, X)$. A natural equivalence relation on $\mathcal{C}(A, X)$ is defined by $f \sim g :\Leftrightarrow \gamma(f) = \gamma(g)$. Then γ induces an injection $\mathcal{C}(A, X)/\!\sim\ \hookrightarrow [A, X]$ by which we identify $\gamma(f)$ with the equivalence class $[f]$. (The *naturality* of $(\sim)$ means that $f \sim g$ implies $ufw \sim ugw$ whenever this makes sense.)

If B is cofibrant, Y fibrant and $f, g \in \mathcal{C}(B, Y)$ then we call f *homotopic to* g ($f \simeq g$) if $\gamma(f) = \gamma(g)$ (equivalently: $[f] = [g]$ or $f \sim g$).

Note that the homotopy relation ($\simeq$) is only defined for maps having the same *cofibrant* source and the same *fibrant* target. In this restricted sense the homotopy relation is a natural equivalence relation.

By (H-3) (ii) the homotopy relation can be described in several ways.

PROPOSITION (1.3)

For cofibrant B and fibrant Y the relations ($\overset{\ell}{\sim}$) *and* ($\overset{r}{\sim}$) *are natural equivalence relations on* $\mathcal{C}(B,Y)$ *and*

$$\begin{aligned} \mathcal{C}(B,Y)/\overset{\ell}{\sim} \;&=\; \mathcal{C}(B,Y)/\simeq \;=\; \mathcal{C}(B,Y)/\overset{r}{\sim} \\ &\overset{\gamma}{\approx}\; [B,Y]\,. \end{aligned}$$

□

We define cylinder objects as sort of "passe-partout" left homotopy objects. This means, *one* cylinder object for B suffices to describe the homotopy relation for *all* maps on B.

Definition (1.4)

Let $B \in \mathcal{C}_c$. A left homotopy object $(\tilde{B}, i_0, i_1, p)$ for B is a *cylinder object* for B if it has the following property: whenever $f,\, g \in \mathcal{C}(B,Y)$ with $Y \in \mathcal{C}_f$ are homotopic, there is a left homotopy $\tilde{B} \to Y$ from f to g.

Path objects are defined in a dual manner.

(H-4) Every cofibrant object has a cylinder object.
Every fibrant object has a path object.

Remarks (1.5)

(a) It would be more standard to call $\operatorname{Ho}\mathcal{C}$ the 'homotopy category' of $\mathcal{C}$. As it is generally accepted to call induced functors on the Ho-level 'derived functors', the terminology used here is more consistent. There is a classical notion of 'derived category' of an abelian category $\mathcal{V}$, i.e. the localization of $\langle \mathrm{DG}^* \rangle_{\mathcal{V}}$ with respect to the class of H^*-equivalences. Although we do not claim that this category is a homotopical category, the ambiguity can hardly cause any trouble.

(b) Every category has a "discrete" homotopy structure with we = isomorphisms. For more important examples of the general kind, see App. A.

N° 2 Left homotopical categories

The axioms for a homotopical category are self-dual, i.e. reversing all arrows and changing the roles of cofibrant and fibrant objects yields a homotopy structure in $\mathcal{C}^{\mathrm{op}}$. Hence each notion and each result has a dual.

The most important results of this chapter, however, make use only of the left homotopy relation and its properties. By omitting the assumptions concerning ($\overset{r}{\sim}$) we obtain weaker axioms (LH-3) and (LH-4), while (H-1) and (H-2) are unchanged.

Definition (1.6)

We call $(\mathcal{C}, we, \mathcal{C}_c, \mathcal{C}_f)$ a *left homotopy structure* in $\mathcal{C}$ if the axioms (H-1), (H-2), (LH-3) and (LH-4) are satisfied. We also call $\mathcal{C}$, endowed with a distinguished left homotopy structure, a *left homotopical category*.

There is a dual notion of *right homotopical category*.

One may note that a homotopical category is the same as a left homotopical category in which every fibrant object has a path object.

N° 3 Homotopical subcategories

Let $\mathcal{C}$ be a homotopical category. If $\mathcal{E} \subset \mathcal{C}$ is a full subcategory, there is an *induced structure*

$$(*) \qquad we_{\mathcal{E}} = we_{\mathcal{C}} \cap \operatorname{mor} \mathcal{E}\,, \quad \mathcal{E}_c = \mathcal{C}_c \cap \mathcal{E}\,, \quad \mathcal{E}_f = \mathcal{C}_f \cap \mathcal{E}\,.$$

Let $\widetilde{\mathcal{E}} \subset \operatorname{Ho}\mathcal{C}$ be the full subcategory with the same objects as $\mathcal{E}$. Let $\gamma_{\mathcal{E}} : \mathcal{E} \to \widetilde{\mathcal{E}}$ be the restriction of γ.

Definitions (1.7)

We call $\mathcal{E} \subset \mathcal{C}$ a *homotopical subcategory* if $\mathcal{E}$ with $(*)$ is a homotopical category, such that the derived category is $\operatorname{Ho}\mathcal{E} = \widetilde{\mathcal{E}}$.

We call $\mathcal{E} \subset \mathcal{C}$ a *localizing subcategory* if $\gamma_{\mathcal{E}} : \mathcal{E} \to \widetilde{\mathcal{E}}$ is a localization of $(\mathcal{E}, we_{\mathcal{E}})$.

We show in § 2, N° 2 that the derived category of a homotopical category is a localization with respect to the class of weak equivalences. Hence a homotopical subcategory is localizing.

LEMMA (1.8)

Let $\mathcal{E}$ with $(*)$ satisfy (H-2) *and* (U). *Then it also satisfies* (H-1), (H-3). *Moreover $\mathcal{E}$ is localizing.*

(U) Let f, $g \in \mathcal{E}(B, Y)$ be homotopic in $\mathcal{C}$ (B cofibrant, Y fibrant).
Then f, g are left (resp. right) homotopic in $\mathcal{E}$.

Proof. Axiom (H-1) is trivially satisfied in $\mathcal{E}$. Let $\operatorname{Ho}\mathcal{E} = \widetilde{\mathcal{E}}$. Then condition (o) of (H-3) is satisfied. For the other conditions let $B \in \mathcal{E}_c$ and $Y \in \mathcal{E}_f$. By the corresponding property of $\mathcal{C}$ we can represent any $\varphi : B \to Y \in \widetilde{\mathcal{E}}$ by a map in $\mathcal{E}$; and if $f, g : B \rightrightarrows Y \in \mathcal{E}$ are such that $\gamma(f) = \gamma(g)$, i.e. $f \simeq g$ in $\mathcal{C}$, then (U) implies $f \overset{\ell}{\sim} g$ (resp. $f \overset{r}{\sim} g$) in $\mathcal{E}$. Hence (H-3) is satisfied. Since the first three axioms imply that $\operatorname{Ho}\mathcal{E}$ is the localization of $(\mathcal{E}, we_{\mathcal{E}})$, it follows by definition that $\mathcal{E}$ is a localizing subcategory. □

THEOREM (1.9)

$\mathcal{E}$ with $(*)$ is a homotopical subcategory iff it satisfies (H-2), (H-4), (U). □

Definition - Remark (1.10)

Analogously, $\mathcal{E}$ is a *left homotopical subcategory* of the (left) homotopical category $\mathcal{C}$, if $\mathcal{E}$ with $(*)$ is a left homotopical category, such that $\operatorname{Ho}\mathcal{E} = \widetilde{\mathcal{E}}$. Note that a left homotopical subcategory is localizing. There are obvious analogs of (1.8) and (1.9) where (U) is replaced by the obvious weaker condition (LU).

There is a dual notion of *right homotopical subcategory*.

COROLLARY (1.11)

Let $\mathcal{C}$ be a homotopical category. Then $\mathcal{C}_c$ is a left and $\mathcal{C}_f$ a right homotopical subcategory.

Moreover, $\operatorname{Ho}\mathcal{C}_c$ *and* $\operatorname{Ho}\mathcal{C}_f$ *are equivalent subcategories of* $\operatorname{Ho}\mathcal{C}$.

Proof. By (1.9) one has to check (H-2), (LH-4), (LU) for $\mathcal{C}_c$ which is trivial. The assertion for $\mathcal{C}_f$ is dual. Since now these are localizing subcategories, the last assertion follows from the fact that every object of $\mathcal{C}$ has a cofibrant (resp. fibrant) model. □

Addendum (1.12)

If cofibrant - fibrant objects have cofibrant - fibrant cylinder objects, then also $\mathcal{C}_f$ and $\mathcal{C}_{cf}$ are left homotopical (thus localizing) subcategories. □

Often $\mathcal{E}$ is characterized by some "homotopy invariant" properties (such as connectivity). To make this precise we define: An object $X \in \mathcal{C}$ is of *$\mathcal{E}$-homotopy type* if there is a finite chain of weak equivalences $X \xleftarrow{\sim} X_1 \xrightarrow{\sim} X_2 \ldots \xleftarrow{\sim} X_n$ with $X_n \in \mathcal{E}$.

Corollary (1.13)

When $\mathcal{E} \subset \mathcal{C}$ contains all objects of $\mathcal{E}$-homotopy type, then $\mathcal{E}$ is a homotopical subcategory.

Proof. By (1.9) this is a matter of checking (H-2), (H-4), (U) for $\mathcal{E}$ which is readily done. □

Let $\mathcal{C}$ be a homotopical category. By (H-2) we have a *choice of cofibrant models* by which we associate to each $X \in \mathcal{C}$ an object $LX \in \mathcal{C}_c$ and a weak equivalence $\ell_X : LX \xrightarrow{\sim} X$. Note that we (may) assume $LX \in \mathcal{C}_{cf}$ for $X \in \mathcal{C}_f$. Dually we have a *choice of fibrant models* in $\mathcal{C}$. By (H-4) we have a *choice of cylinder objects* (one for each cofibrant object).

Definition (1.14)

Let $\mathcal{E} \subset \mathcal{C}$ be a full subcategory endowed with the induced structure. Suppose the embedding has a right adjoint - left inverse $E : \mathcal{C} \to \mathcal{E}$. We call E an *Eilenberg functor* if $E(we_f) \subset we_f$ and $j_Y : E(Y) \to Y$ is a weak equivalence if Y is fibrant and of $\mathcal{E}$-homotopy type.

We recall some facts on 'reflective subcategories' [Mc]. When E is right adjoint, it is also left inverse to to the (full) embedding up to natural isomorphism. By definition the adjunction map $j_X : E(X) \to X \in \mathcal{C}$ is universal for maps from $\mathcal{E}$-objects to X. Hence we may assume that E is left inverse, and that $j_X = 1$ for $X \in \mathcal{E}$.

The last condition of (1.14) follows if for any such Y there is a weak equivalence $\rho : Y \xrightarrow{\sim} Y'$ with $Y' \in \mathcal{E}_f$. In fact, $E(\rho) = \rho j_X$ and thus j_X is a weak equivalence by the first condition.

Proposition (1.15)

Let $\mathcal{C}$ be a right homotopical category. Under the following conditions $\mathcal{E} \subset \mathcal{C}$ is a right homotopical subcategory.

(i) For some choice of cofibrant models in $\mathcal{C} : X \in \mathcal{E} \Rightarrow LX \in \mathcal{E}$.

(ii) There is an Eilenberg functor $E : \mathcal{C} \to \mathcal{E}$.

Proof. To verify (H-2) for $\mathcal{E}$ let $A \in \mathcal{E}$ be given. Let $r : A \xrightarrow{\sim} RA$ be a fibrant model in $\mathcal{C}$, then since j_{RA} is a weak equivalence, so is $E(r) : A \xrightarrow{\sim} E(RA) = R'A \in \mathcal{E}$ and $R'A$ is fibrant. By (i) there is a cofibrant - fibrant model $M'A \xrightarrow{\sim} R'A$ in $\mathcal{E}$, and if A is cofibrant there is a lifting $A \xrightarrow{\sim} M'A$; cf. (1.18). Hence we have a choice of fibrant models in $\mathcal{E}$. By (i) we also have a choice of cofibrant models. To prove

(RU) let $f, g : B \rightrightarrows Y \in \mathcal{E}$ (Y fibrant). When $f \overset{r}{\sim} g$ in $\mathcal{C}$ via the right homotopy $H : B \to \tilde{Y}$, then $E(\tilde{Y})$ is a right homotopy object and $E(H) : B \to E(\tilde{Y})$ a right homotopy $f \overset{r}{\sim} g$ in $\mathcal{E}$. The verification of (RH-4) is similar: when $\tilde{Y}$ is a path object for Y in $\mathcal{C}$, then $E(\tilde{Y})$ is such one in $\mathcal{E}$. □

There is a more simple criterion which we formulate in the dual situation.

PROPOSITION (1.16)

Let $\mathcal{C}$ be a left homotopical category. Under the following conditions $\mathcal{E} \subset \mathcal{C}$ is a left homotopical subcategory.

(iii) For some choice of cylinder objects in $\mathcal{C}$: $B \in \mathcal{E} \Rightarrow \tilde{B} \in \mathcal{E}$.

(iv) For some choice of cofibrant and of fibrant models in $\mathcal{C}$: $X \in \mathcal{E} \Rightarrow LX, RX \in \mathcal{E}$.

Proof. Plainly (iv) implies (H-2) and (iii) implies (LH-4) (for $\mathcal{E}$). To verify (LU) let $f \simeq g$ in $\mathcal{C}$, then for any cylinder object $\tilde{B}$ there is a left homotopy $\tilde{B} \to Y$. In case $B \in \mathcal{E}$ we can choose $\tilde{B} \in \mathcal{E}$ by (iii), thus $f \overset{\ell}{\sim} g$ in $\mathcal{E}$ as desired. □

THEOREM (1.17)

Let $\mathcal{C}$ be a homotopical category. Under the conditions (i)–(iii) *above, $\mathcal{E} \subset \mathcal{C}$ is a homotopical subcategory.*

Proof. By the propositions above, $\mathcal{E}$ is a right homotopical subcategory satisfying (LH-4). This is the assertion, cf. (1.6). □

§ 2 FUNDAMENTAL RESULTS

Let $\mathcal{C}$ be a homotopical category.

N° 1 Lifting and extension

We begin with some immediate consequences of the axioms.

LIFTING LEMMA (1.18)

Let $f : B \to Y \overset{\sim}{\leftarrow} Z : p$ be maps in $\mathcal{C}$ with $p \in we$, B cofibrant and Y, Z fibrant. Then there is $g : B \to Z$ with $pg \simeq f$ and g is unique up to homotopy. □

EXTENSION LEMMA (1.19)

Let $f : B \to Y$, $j : B \overset{\sim}{\to} C$ be maps in $\mathcal{C}$ with $j \in we$ and B, C cofibrant, Y fibrant. Then there is $g : C \to Y$ with $gj \simeq f$ and g is unique up to homotopy. □

COROLLARY (1.20)

Let both B and Y be cofibrant-fibrant and let $f : B \overset{\sim}{\to} Y$ be a weak equivalence. Then f is a homotopy equivalence, i.e. there is $g : Y \to B$ such that $fg \simeq 1$ and $gf \simeq 1$. □

These are easy consequences of (H-3). The following is a variant of (H-2).

Lemma (1.21)

For each object X there is a commutative (up to homotopy) diagram in $\mathcal{C}$, with LX cofibrant, RX fibrant and MX cofibrant-fibrant:

$$\begin{array}{ccccc} & & X & & \\ & \overset{\ell_X}{\underset{\sim}{\nearrow}} & & \overset{r_X}{\underset{\sim}{\searrow}} & \\ LX & & & & RX \\ & \underset{r'_X}{\overset{\sim}{\searrow}} & & \underset{\ell'_X}{\overset{\sim}{\nearrow}} & \\ & & MX & & \end{array}$$

We choose such a *diagram of models* for each X. In case X cofibrant we choose $LX = X$ and $RX = MX$, etc.

Proof. Put $MX = RLX$ according to (H-2). Then ℓ'_X exists by (1.19) and is a weak equivalence, since by (H-1) if $f \overset{\ell}{\sim} g$ and f is a weak equivalence so is g. □

N° 2 The derived category

The next lemma is an easy consequence of (H-1).

Lemma (1.22)

If $H : \mathcal{C} \to \mathcal{K}$ is a functor that carries weak equivalences to isomorphisms, then left (or right) homotopic maps in $\mathcal{C}$ are carried to the same map in $\mathcal{K}$. □

Theorem (1.23)

The derived category $\gamma : \mathcal{C} \to \mathrm{Ho}\,\mathcal{C}$ is a localization of $(\mathcal{C}, we)$.

Proof. By (H-3) γ carries weak equivalences to isomorphisms. We have to show: if a functor $H : \mathcal{C} \to \mathcal{K}$ does the same, there is unique $\bar{H} : \mathrm{Ho}\,\mathcal{C} \to \mathcal{K}$ with $\bar{H}\gamma = H$. On objects put $\bar{H}(X) = H(X)$. In order to define $\bar{H}$ on morphisms choose for each object $X \in \mathcal{C}$ a diagram of models (1.21). Let $\varphi \in [X, Y]$ be given. By (H-3) we find unique (up to homotopy) $f \in \mathcal{C}(MX, MY)$ with $\varphi = \gamma(\ell_Y)\gamma(r'_Y)^{-1}\gamma(f)\gamma(r'_X)\gamma(\ell_X)^{-1}$.

Now put $\bar{H}(\varphi) = H(\ell_Y)H(r'_Y)^{-1}H(f)H(r'_X)H(\ell_X)^{-1}$ which is well-defined by (1.22). Then it is easy to see that $\bar{H}$ is a functor. We also have $\bar{H}(\varphi) = H(r_Y)^{-1}H(f')H(\ell_X)^{-1}$ for any $f' \in \mathcal{C}(LX, RY)$ with $\varphi = \gamma(r_Y)^{-1}\gamma(f')\gamma(\ell_X)^{-1}$. It easily follows that $\bar{H}(\gamma(g)) = H(g)$ for $g \in \mathcal{C}(X, Y)$, thus $\bar{H}$ is as desired. Clearly $\bar{H}$ had to be defined this way. □

Remark (1.24)

Suppose there exists a functor $H : \mathcal{C} \to \mathcal{K}$ with the property: $f \in we \Leftrightarrow H(f)$ is an isomorphism. This is practically always the case. By (1.23) this implies and follows from the property: $f \in we \Leftrightarrow \gamma(f)$ is an isomorphism. In this case the class of weak equivalences is thus determined by the derived category $\mathrm{Ho}\,\mathcal{C}$ (and conversely). In particular, the weak equivalences in $\mathcal{C}_{cf}$ are *precisely* the homotopy equivalences.

Theorem (1.25)

(a) *Two maps $f, g \in \mathcal{C}(A, X)$ are equivalent, $\gamma(f) = \gamma(g)$, iff $rf\ell \simeq rg\ell$ for some weak equivalences r and ℓ.*

(b) *Every morphism in $\mathrm{Ho}\,\mathcal{C}$ is of the form $\gamma(r)^{-1}\gamma(f)\gamma(\ell)^{-1}$ where r, f, ℓ are maps in $\mathcal{C}$ with $r, \ell \in we$.*

(c) *Let $\mathcal{C}_{cf}/_{\simeq}$ be the category obtained from $\mathcal{C}_{cf}$ by identifying homotopic maps. The functor γ induces an equivalence of categories*

$$\mathcal{C}_{cf}/_{\simeq} \overset{\approx}{\hookrightarrow} \operatorname{Ho}\mathcal{C}\,.$$

□

THEOREM (1.26)

Let $\Phi_1, \Phi_2 : \operatorname{Ho}\mathcal{C} \rightrightarrows \mathcal{K}$ be functors, and $\alpha : \Phi_1\gamma \to \Phi_2\gamma$ a natural map. Then so is $\bar{\alpha} : \Phi_1 \to \Phi_2$ defined by $\bar{\alpha}_X = \alpha_X$.

It is then easy to see that $\alpha \mapsto \bar{\alpha}$ is a bijection $\operatorname{Hom}(\Phi_1\gamma, \Phi_2\gamma) = \operatorname{Hom}(\Phi_1, \Phi_2)$ where Hom denotes the class of natural maps. The inverse mapping is $\bar{\alpha} \mapsto \bar{\alpha} * \gamma$.

Proof. Recall that $\gamma(X) = X$ on objects and that every morphism in $\operatorname{Ho}\mathcal{C}$ is a finite composition of maps $\gamma(f)$ or $\gamma(f)^{-1}$. This implies the assertion, since clearly $\bar{\alpha}$ is natural with respect to every map in the image of γ. □

N° 3 Homotopical functors and their derived functors

Let $\mathcal{C}$ and $\mathcal{D}$ be homotopical categories.

Fully homotopical functor — fully derived functor (1.27)

Let $F : \mathcal{C} \to \mathcal{D}$ be a functor. We call F a *fully homotopical functor* if $F(we) \subset we$, i.e. if F preserves weak equivalences. In this case, by (1.23), there is unique $\widetilde{F} : \operatorname{Ho}\mathcal{C} \to \operatorname{Ho}\mathcal{D}$ satisfying $\widetilde{F}\gamma = \gamma F$, i.e. for all objects X and maps f in $\mathcal{C}$

$$(*) \qquad \widetilde{F}(X) = F(X)\,, \qquad \widetilde{F}([f]) = [F(f)]\,.$$

We call $\widetilde{F}$ (if existent) the *fully derived functor* of F.

Given homotopical functors $F_1, F_2 : \mathcal{C} \rightrightarrows \mathcal{D}$ and a natural map $v : F_1 \to F_2$, we have the induced (or *derived*) natural map $[v] : \widetilde{F}_1 \to \widetilde{F}_2$ defined by $[v]_X = [v_X]$, cf. (1.26).

Left homotopical functor — left derived functor (1.28)

We call F a *left homotopical functor* if $F(we_c) \subset we_c$. Note that this comprises the conditions $F(\mathcal{C}_c) \subset \mathcal{D}_c$ and $F(we_c) \subset we$. (The former is of importance in N° 4, but here it is not needed. It implies that a composition of left homotopical functors is again such. But because of this condition a fully homotopical functor need not be left homotopical.)

Let F be left homotopical; or more generally (1.11), suppose there is a full subcategory $\mathcal{E} \subset \mathcal{C}$, such that (i) $\mathcal{E}$ is localizing, (ii) for all $X \in \mathcal{C}$ there is a weak equivalence $\varepsilon_X : EX \overset{\sim}{\to} X$ with $EX \in \mathcal{E}$, (iii) $F(we_{\mathcal{E}}) \subset we$. — Then there is unique $\widetilde{F} : \operatorname{Ho}\mathcal{E} \to \operatorname{Ho}\mathcal{D}$ satisfying $\widetilde{F}\gamma\,|_{\mathcal{E}} = \gamma F\,|_{\mathcal{E}}$, i.e. $(*)$ holds for objects and maps in $\mathcal{E}$. We define a functor F^{L} as the composite

$$F^{\mathsf{L}} : \operatorname{Ho}\mathcal{C} \underset{\sim}{\overset{E}{\longrightarrow}} \operatorname{Ho}\mathcal{E} \overset{\widetilde{F}}{\longrightarrow} \operatorname{Ho}\mathcal{D}\,.$$

By (i)–(ii) we get the functor E which is left inverse to the embedding ($\varepsilon_X = \mathrm{id}$ if $X \in \mathcal{E}$), and an equivalence of categories. It is constructed in such a way that $\{[\varepsilon_X] | X \in \mathcal{C}\}$ is a *natural* isomorphism $\operatorname{Emb} \circ E \cong \operatorname{Id}$.

We call F^{L} (or any isomorphic functor) a *left derived* of F. It follows from (1.29) that F^{L} is uniquely determined by F (up to canonical isomorphism).

PROPOSITION (1.29)

The functor F^{L} *together with the natural map* $\lambda : F^{\mathsf{L}}\gamma \to \gamma F$, $\lambda_X = \gamma F(\varepsilon_X)$, *has the following universal property:*

Given $\Phi : \mathrm{Ho}\,\mathcal{C} \to \mathrm{Ho}\,\mathcal{D}$ *with* $\zeta : \Phi\gamma \to \gamma F$ *there is unique* $\bar{\zeta} : \Phi \to F^{\mathsf{L}}$ *satisfying* $\lambda(\bar{\zeta} * \gamma) = \zeta$. □

The proof is omitted, cf. [Q2|I.4]. We note the following useful consequence: Given left homotopical functors $F_i : \mathcal{C} \to \mathcal{D}$ (with $\lambda_i : F_i^{\mathsf{L}}\gamma \to \gamma F_i$), $i \in \{0,1\}$, and a natural map $v : F_1 \to F_2$, there is unique natural map $v^{\mathsf{L}} : F_1^{\mathsf{L}} \to F_2^{\mathsf{L}}$ satisfying $\lambda_2(v^{\mathsf{L}} * \gamma) = (\gamma * v)\lambda_1$.

The larger we can choose $\mathcal{E}$, the stronger is the relationship between F and F^{L}. In fact, by construction (1.27) (∗) holds for all objects and maps in $\mathcal{E}$. In particular, if F is fully homotopical, then we may choose $\mathcal{E} = \mathcal{C}$ and $F^{\mathsf{L}} = \widetilde{F}$.

There are dual notions: a *right homotopical functor* F satisfies $F(\mathcal{C}_f) \subset \mathcal{D}_f$, $F(we_f) \subset we$. The latter condition implies that F has a *right derived functor* F^{R}.

We call a full subcategory $\mathcal{M} \subset \mathcal{C}_{cf}$ *sufficiently large* if every object of $\mathcal{C}$ has a model in $\mathcal{M}$. Equivalently, every object of $\mathcal{C}_{cf}$ is homotopy equivalent to some object of $\mathcal{M}$. Again equivalently, the embedding $\mathcal{M}/_{\simeq} \hookrightarrow \mathrm{Ho}\,\mathcal{C}$ is an equivalence of categories, cf. (1.25) (c).

Homotopical functor — derived functor (1.30)

Generalizing all the notions introduced above, we say F is a *homotopical functor* if there is a sufficiently large subcategory $\mathcal{M} \subset \mathcal{C}_{cf}$ and a functor $\widetilde{F} : \mathcal{M}/_{\simeq} \to \mathrm{Ho}\,\mathcal{D}$ satisfying (∗) for all objects and maps in $\mathcal{M}$. Clearly $\widetilde{F}$ is unique, it exists iff F carries homotopic maps in $\mathcal{M}$ to equivalent maps. — If F is homotopical, there is a functor $F' : \mathrm{Ho}\,\mathcal{C} \to \mathrm{Ho}\,\mathcal{D}$ (essentially unique) which agrees with $\widetilde{F}$ on $\mathcal{M}/_{\simeq}$. We call F' (or any isomorphic functor) a *derived functor* of F.

Notice that any of $\widetilde{F}, F^{\mathsf{L}}, F^{\mathsf{R}}$ (if existent) is a derived functor of F.

All this works as well for *left* homotopical categories, except that for a right homotopical functor we may not be able to construct a right derived functor. For this we need that $\mathcal{C}_f$ is localizing; cf. (1.12).

N° 4 The Adjoint Functor Theorem

Let $\mathcal{C}$ *and* $\mathcal{D}$ *be homotopical categories.*

Let $F : \mathcal{C} \rightleftarrows \mathcal{D} : G$ be adjoint functors, with adjunction maps $u : FG \to \mathrm{Id}$, $v : \mathrm{Id} \to GF$. If both F and G are fully homotopical, then $\widetilde{F} : \mathrm{Ho}\,\mathcal{C} \rightleftarrows \mathrm{Ho}\,\mathcal{D} : \widetilde{G}$ are also adjoint, with adjunction maps $[u]$, $[v]$, cf. (1.27). Moreover, if u, v are natural weak equivalences, then $[u]$, $[v]$ are natural isomorphisms and thus $\widetilde{F}$, $\widetilde{G}$ are adjoint equivalences of categories.

We prove a variant of this result for left/right homotopical functors.

Definitions (1.31)

A functor $F : \mathcal{C} \to \mathcal{D}$ is a *homotopical equivalence* if it is a homotopical functor and $F' : \mathrm{Ho}\,\mathcal{C} \to \mathrm{Ho}\,\mathcal{D}$ an equivalence of categories. — Let G be right adjoint to F. Then F, G are *adjoint homotopical equivalences* if these are homotopical functors, such that the derived functors are adjoint equivalences of categories.

Let $F : \mathcal{C} \rightleftarrows \mathcal{D} : G$ be adjoint functors. Then (F, G) is an *adjoint homotopical pair* if F is a left homotopical functor and G a right homotopical functor.

ADJOINT FUNCTOR THEOREM (1.32)

Let (F, G) be an adjoint homotopical pair.

(a) *Then also the derived functors F^{L}, G^{R} are adjoint.*

(b) *The derived functors are adjoint equivalences if the following condition is satisfied.*

(E) (i) F preserves and reflects weak equivalences.
(ii) $u_V : FG(V) \xrightarrow{\sim} V$ is a weak equivalence if $V \in \mathcal{D}_f$.

Here $u_V = 1^{\#}_{G(V)}$, the adjunction map in $\mathcal{D}$. Note that (E) together with the conditions $F(\mathcal{C}_c) \subset \mathcal{D}_c$ and $G(\mathcal{D}_f) \subset \mathcal{C}_f$ imply that (F, G) is a homotopical pair.

Proof. *Ad* (a). Recall F, G adjoint means there is a natural bijection

$$\# : \mathcal{C}(A, G(U)) \approx \mathcal{D}(F(A), U) : \flat$$

under $\mathcal{C}^{\mathrm{op}} \times \mathcal{D}$. Let $B \in \mathcal{C}_c$, $V \in \mathcal{D}_f$ and $f,\, g : B \to G(V)$. If $f \simeq g$, there is a left homotopy H from f to g defined on a left homotopy object $\tilde{B}$. Since $\tilde{B}$ is cofibrant, $F(\tilde{B})$ is a left homotopy object for $F(B)$ and, by naturality, $H^{\#}$ is a left homotopy $f^{\#} \simeq g^{\#}$. In a dual manner, $f^{\#} \simeq g^{\#}$ implies $f \simeq g$. Whence the bijection

$$\# \;:\; [B, G(V)] \approx [F(B), V] \;:\; \flat \qquad (B \in \mathcal{C}_c\,,\; V \in \mathcal{D}_f) \tag{1.33}$$

which is clearly natural under $(\mathcal{C}_c)^{\mathrm{op}} \times \mathcal{D}_f$ hence (with F, G replaced by $\widetilde{F}, \widetilde{G}$) under $(\mathrm{Ho}\,\mathcal{C}_c)^{\mathrm{op}} \times \mathrm{Ho}\,\mathcal{D}_f$, cf. proof of (1.26). It easily follows that the composite bijection

$$[A, G^{\mathsf{R}}(U)] \underset{\approx}{\xrightarrow{[\ell_A]^*}} [LA, \widetilde{G}(RU)] \underset{\approx}{\xrightarrow{\#}} [\widetilde{F}(LA), RU] \underset{\approx}{\xleftarrow{[r_U]_*}} [F^{\mathsf{L}}(A), U] \tag{1.34}$$

is natural under $(\mathrm{Ho}\,\mathcal{C})^{\mathrm{op}} \times \mathrm{Ho}\,\mathcal{D}$ proving the assertion.

Ad (b). It is easy to see that (E) implies

$$f : B \to G(V) \in we \;\Leftrightarrow\; f^{\#} : F(B) \to V \in we \qquad (B \in \mathcal{C}_c\,,\; V \in \mathcal{D}_f)\,. \tag{E$'$}$$

This condition in turn implies that the adjunction maps for F^{L}, G^{R} given by (1.34) are isomorphisms. □

Remarks (1.35)

(a) When F, G is an adjoint homotopical pair, it follows from (1.33) that if $\tilde{B}$ is a cylinder object for $B \in \mathcal{C}_c$ then so is $F(\tilde{B})$ for $F(B)$. Dually for G.

(b) In statement and proof of (1.32) it suffices that G is defined on $\mathcal{D}_f$ (resp. that F is defined on $\mathcal{C}_c$).

(c) In (1.32) it suffices that $\mathcal{C}$ is a *left* and $\mathcal{D}$ a *right* homotopical category. Hence it holds, for instance, if $\mathcal{C}$ and $\mathcal{D}$ are cofibration or fibration categories (see App. A). Notice that we make no use of cylinder or path objects.

§ 3 Comonoids Up To Homotopy

Let $\mathcal{C}$ be a homotopical category.

N° 1 ... as comonoids over the derived category
Let us assume that $\mathrm{Ho}\,\mathcal{C}$ is given the structure of a monoidal category, i.e. $\mathrm{Ho}\,\mathcal{C}$ is endowed with a monoidal structure $\otimes, K, \bar{\alpha}, \bar{\lambda}, \bar{\rho}, \bar{\tau}$, cf. App. B, N° 1. In N° 2 and N° 3 we give conditions for a monoidal structure of $\mathcal{C}$ to induce a monoidal structure of $\mathrm{Ho}\,\mathcal{C}$.

Let cCom-$\mathrm{Ho}\,\mathcal{C}$ be the category of *cocommutative comonoids* over $\mathrm{Ho}\,\mathcal{C}$, i.e. of triples $(A, \Psi, \mathfrak{e})$ with A and $\Psi : A \to A \otimes A$ and $\mathfrak{e} : A \to K$ in $\mathrm{Ho}\,\mathcal{C}$ such that

$$(1.36) \qquad \pi_1 \Psi = 1_A = \pi_2 \Psi \ , \quad (\Psi \otimes 1_A)\Psi = \bar{\alpha}(1_A \otimes \Psi)\Psi \ , \quad \bar{\tau}\Psi = \Psi \, .$$

Here we use the natural maps $\pi_1 = \bar{\rho}(1_A \otimes \mathfrak{e})$, $\pi_2 = \bar{\lambda}(\mathfrak{e} \otimes 1_A) : A \otimes A \to A$. Note that a *map* of comonoids (over $\mathrm{Ho}\,\mathcal{C}$) $\zeta : (A, \Psi, \mathfrak{e}) \to (B, \Psi', \mathfrak{e}')$ is a map in $\mathrm{Ho}\,\mathcal{C}$ satisfying

$$(1.37) \qquad \Psi'\zeta = (\zeta \otimes \zeta)\Psi \quad , \quad \mathfrak{e}'\zeta = \mathfrak{e} \, .$$

Remark (1.38)
When K is fibrant and a terminal object of $\mathcal{C}$, then K is also terminal in $\mathrm{Ho}\,\mathcal{C}$. Hence there is unique $\mathfrak{e} : A \to K \in \mathrm{Ho}\,\mathcal{C}$ and the second condition in (1.37) is automatic. In fact, $\mathfrak{e} = [\varepsilon]$ where $\varepsilon : A \to K$ is the unique map.

Definition (1.39)
Define a category cComh $\mathcal{C}$ as follows: the objects are the same as in cCom-$\mathrm{Ho}\,\mathcal{C}$, but a *morphism* $f : (A, \Psi, \mathfrak{e}) \to (B, \Psi', \mathfrak{e}')$ in cComh $\mathcal{C}$ is a map in $\mathcal{C}$ such that $[f] : A \to B \in \mathrm{Ho}\,\mathcal{C}$ is a morphism of comonoids. An object $A = (A, \Psi, \mathfrak{e})$ in cComh $\mathcal{C}$ is called a *cocommutative comonoid up to homotopy* (*cComh*) over $\mathcal{C}$; and Ψ is called the *homotopy diagonal*, $\mathfrak{e}$ the *homotopy counit* of A.

Let γ' : cComh $\mathcal{C} \to$ cCom-$\mathrm{Ho}\,\mathcal{C}$ be the functor $A \mapsto A$, $f \mapsto [f]$. We have a commutative diagram

$$(1.40) \qquad \begin{array}{ccc} \mathrm{cComh}\,\mathcal{C} & \xrightarrow{\gamma'} & \mathrm{cCom\text{-}Ho}\,\mathcal{C} \\ {\scriptstyle \phi}\downarrow & & \downarrow{\scriptstyle \phi} \\ \mathcal{C} & \xrightarrow{\gamma} & \mathrm{Ho}\,\mathcal{C} \end{array}$$

where each ϕ is a forgetful functor. Obviously these are faithful and reflect isomorphisms: if $\phi(f)$ is an isomorphism so is f. One may observe that (1.40) is a pull-back of categories.

We define: A map in cComh $\mathcal{C}$ is a *weak equivalence*, and an object of cComh $\mathcal{C}$ is *cofibrant* (resp. *fibrant*), if it is so-called in $\mathcal{C}$.

THEOREM (1.41)

With this structure cComh$\mathcal{C}$ *is a homotopical category, and* cCom-Ho$\mathcal{C}$ *(with* γ'*) its derived category.*

Two maps in cComh$\mathcal{C}$ *are left (resp. right) homotopic iff they are so in* $\mathcal{C}$.

Proof. (H-1) is clear and lemmae (1.42)–(1.43) easily imply (H-2), (H-4) and the last assertion of the theorem. For instance, a cofibrant model for $(A, \Psi, \mathfrak{e})$ is obtained by lifting Ψ and $\mathfrak{e}$ to LA (in Ho$\mathcal{C}$). — *Ad* (H-3). Since the right-hand ϕ in (1.40) reflects isomorphisms, γ' carries weak equivalences to isomorphisms. Now let $B, Y \in$ cComh$\mathcal{C}$ with B cofibrant, Y fibrant. Then γ' : cComh$\mathcal{C}(B, Y) \to$ cCom-Ho$\mathcal{C}(B, Y)$ is surjective, since any φ in the range is of the form $\gamma(f)$ and then, by definition, f is in the domain. (Note that the mapping γ' is the restriction of γ). Finally, since ϕ on the left of (1.40) is faithful, the last assertion of the theorem shows that $\gamma'(f) = \gamma'(g)$ iff $f \overset{\ell}{\sim} g$ (resp. $f \overset{r}{\sim} g$). □

We need here the following trivial facts.

LEMMA (1.42)

(a) *Let* $s : B \xrightarrow{\sim} A$ *(or* $s : A \xrightarrow{\sim} B$ *)* $\in \mathcal{C}$ *be a weak equivalence and* $A \in$ cComh$\mathcal{C}$. *Then there is a unique homotopy diagonal and a unique homotopy counit for* B, *such that* B *and* s *lie in* cComh$\mathcal{C}$.

(b) *Let* $s : B \xrightarrow{\sim} A$ *be a weak equivalence and* C *an object of* cComh$\mathcal{C}$, *further* $g : C \to B$ *(resp.* $g : A \to C$ *) a map in* $\mathcal{C}$. *Then* g *is a morphism in* cComh$\mathcal{C}$ *iff* $s \circ g$ *(resp.* $g \circ s$ *) is so.* □

LEMMA (1.43)

Let $B \in$ cComh$\mathcal{C}$ *be cofibrant, and let* $\tilde{B} = (\tilde{B}, i_0, i_1, q)$ *be a left homotopy object in* $\mathcal{C}$.

(a) *For a unique homotopy diagonal and unique homotopy counit on* $\tilde{B}$ *the maps* i_0, i_1 *and* q *are morphisms in* cComh$\mathcal{C}$. *Hence* $\tilde{B}$ *is a left homotopy object for* B *in* cComh$\mathcal{C}$.

(b) *Let* $f, g : B \rightrightarrows X \in$ cComh$\mathcal{C}$, *and* $H : \tilde{B} \to X$ *a homotopy from* f *to* g *in* $\mathcal{C}$. *Then* H *is a homotopy from* f *to* g *in* cComh$\mathcal{C}$.

(c) *If* $\tilde{B}$ *is a cylinder object for* B *in* $\mathcal{C}$, *then the same holds in* cComh$\mathcal{C}$. □

N° 2 Derived tensor product

Of course there is usually given monoidal structure in $\mathcal{C}$ rather than Ho$\mathcal{C}$. Hence we need conditions under which "derived" monoidal structure is obtained. We give three such conditions, starting with the most convenient. The other conditions are discussed in N° 3. In any case, assuming all objects fibrant, there is a notion of "individual" comonoid up to homotopy, which is more obvious than — and essentially equivalent to — that of cofibrant comonoid up to homotopy.

LEMMA (1.44)

If $\mathcal{A}$ *and* $\mathcal{B}$ *are homotopical categories, then so is* $\mathcal{A} \times \mathcal{B}$ *with the obvious structure* $(we \times we, \mathcal{A}_c \times \mathcal{B}_c, \mathcal{A}_f \times \mathcal{B}_f)$. *The derived category is* Ho$\mathcal{A} \times$ Ho$\mathcal{B}$. □

This is straightforward. Using (1.23) we infer that $\gamma \times \gamma : \mathcal{A} \times \mathcal{B} \to$ Ho$\mathcal{A} \times$ Ho$\mathcal{B}$ is the localization with respect to $we \times we$. (Actually localization is compatible with finite products in general, but this is not needed).

Let the homotopical category $\mathcal{C}$ be endowed with monoidal structure $\otimes$, K, $\alpha, \lambda, \rho, \tau$. We assume that these structures are compatible in the sense that the following condition is satisfied:

(T) If f and g are weak equivalences in $\mathcal{C}$ so is $f \otimes g$,
i.e. $we \otimes we \subset we$.

Note that (T) demands that $\otimes : \mathcal{C} \times \mathcal{C} \to \mathcal{C}$ be a homotopical functor. Hence by (1.44) there is a unique functor $\widetilde{\otimes} : \mathrm{Ho}\,\mathcal{C} \times \mathrm{Ho}\,\mathcal{C} \to \mathrm{Ho}\,\mathcal{C}$ with $\widetilde{\otimes} \circ (\gamma \times \gamma) = \gamma \circ \otimes$ which means $A \widetilde{\otimes} B = A \otimes B$ and $[f] \widetilde{\otimes} [g] = [f \otimes g]$. Henceforth we write $\otimes$ also for the fully derived functor $\widetilde{\otimes}$. Thus we have

$$(1.45) \qquad A \otimes B = A \otimes B\,, \qquad [f] \otimes [g] = [f \otimes g]$$

where on the left (resp. right) of each equation the $\otimes$-product happens in $\mathrm{Ho}\,\mathcal{C}$ (resp. $\mathcal{C}$). Now define $[\alpha]_{A,B,C} = [\alpha_{A,B,C}] : A \otimes (B \otimes C) \cong (A \otimes B) \otimes C \in \mathrm{Ho}\,\mathcal{C}$, then $[\alpha]$ is a natural isomorphism over $\mathrm{Ho}\,\mathcal{C} \times \mathrm{Ho}\,\mathcal{C} \times \mathrm{Ho}\,\mathcal{C}$. In the same way define $[\lambda]$, $[\rho]$ and $[\tau]$; cf. (1.27). Then the following is obvious.

Theorem (1.46)
The category $\mathrm{Ho}\,\mathcal{C}$ *is monoidal, with the structure* $\otimes$, K, $[\alpha], [\lambda], [\rho], [\tau]$. □

We assume now that *all objects of* $\mathcal{C}$ *are fibrant.*

Definition (1.47)
An *individual cComh* is an object $A \in \mathcal{C}_c$ with maps $\psi : A \to A \otimes A$, $\varepsilon : A \to K \in \mathcal{C}$ such that all identities for (A, ψ, ε) to be a cocommutative comonoid (over $\mathcal{C}$) hold up to homotopy, i.e. $\pi_1\psi \simeq 1_A \simeq \pi_2\psi$, etc. A *map* $f : A \to B$ of individual cComhs is a map in $\mathcal{C}$ satisfying $\psi_B f \simeq (f \otimes f)\psi_A$ and $\varepsilon_A f \simeq \varepsilon_B$. The individual cComhs form thus a category denoted by ***cComh*** $\mathcal{C}$.

Let **cComh** $\mathcal{C} \subset$ cComh $\mathcal{C}$ be the subcategory of cofibrant (- fibrant) objects. Recall that this is at least a left homotopical subcategory, cf. (1.11), (1.12).

Theorem (1.48)
The obvious functor q : ***cComh*** $\mathcal{C} \to$ **cComh** $\mathcal{C}$ *is an equivalence of categories.* Hence: ***cComh*** $\mathcal{C}$ *has a canonical left homotopy structure.* □

This functor sends (A, ψ, ε) to $(A, [\psi], [\varepsilon])$ and f to f. It is fully faithful and surjective on objects, hence an equivalence. We may regard q as the functor which identifies two individual cComhs (A, ψ, ε) and $(A', \psi', \varepsilon')$ iff $A = A'$, $\psi \simeq \psi'$, $\varepsilon \simeq \varepsilon'$. Note that the identity map is an isomorphism between such individual cComhs.

It is thus clear that ***cComh*** $\mathcal{C}$ has a canonical left homotopy structure (all objects are cofibrant - fibrant), since this holds for the equivalent category **cComh** $\mathcal{C}$.

Example. For a suitable choice of $\mathcal{C}$ etc. the individual cComhs are precisely the (well-pointed) homotopy-coassociative and -commutative co-H-spaces in topology.

N° 3 Generalizations

There are useful examples of homotopical categories $\mathcal{C}$ with monoidal structure, such that (T) is not satisfied, but where one expects similar results.

Consider the following condition:

(T′) There is a localizing subcategory $\mathcal{E} \subset \mathcal{C}$ containing $\mathcal{C}_c$ and K. If f, g are weak equivalences in $\mathcal{E}$ so is $f \otimes g$, i.e. $\mathcal{E} \otimes \mathcal{E} \subset \mathcal{E}$, $we_{\mathcal{E}} \otimes we_{\mathcal{E}} \subset we$.

For our purposes we may assume in addition that $\mathcal{E}$ is a left homotopical subcategory of $\mathcal{C}$, which implies it is localizing. (Analizing the arguments below one finds that indeed the latter is sufficient). Then $\mathcal{E}$ is left homotopical, monoidal and satisfies (T). Hence we can apply to it the results of N° 2.

It is clear that the monoidal structure of $\operatorname{Ho}\mathcal{E}$ $(\otimes, K, [\alpha], \dots)$ can be extended (essentially uniquely) to a monoidal structure of $\operatorname{Ho}\mathcal{C}$ denoted by $(\bar{\otimes}, K, \bar{\alpha}, \dots)$. By definition $\bar{\otimes}$ is the composite

$$\bar{\otimes} : \operatorname{Ho}\mathcal{C}^{\times 2} \underset{\sim}{\xrightarrow{E^{\times 2}}} \operatorname{Ho}\mathcal{E}^{\times 2} \xrightarrow{\otimes} \operatorname{Ho}\mathcal{E} \underset{\sim}{\hookrightarrow} \operatorname{Ho}\mathcal{C}\,.$$

Hence for any pair of objects of $\mathcal{C}$ we have $X \bar{\otimes} Y = EX \otimes EY$. Here E is left inverse to the embedding $\operatorname{Ho}\mathcal{E} \hookrightarrow \operatorname{Ho}\mathcal{C}$. It is defined as in (1.28). Notice that by definition $\bar{\otimes}$ is a left derived of $\otimes : \mathcal{C}^{\times 2} \to \mathcal{C}$.

Assuming all objects fibrant, we consider the individual cComhs over $\mathcal{C}$ and observe that we get the same if we work over $\mathcal{E}$ (since the latter contains all cofibrant objects).

Summarizing we see that condition (T′) is sufficient to obtain the results of N° 2— with the exception of course that (1.45) only holds for objects and maps in $\mathcal{E}$.

Example (1.49)

Consider the homotopical, monoidal category $\langle\text{dgA}\rangle$, cf. Chp. 4, § 1, N° 2. Unless the ground ring is a field, (T) is not satisfied, but (T′) holds with $\mathcal{E}$ = the left homotopical subcategory of degreewise flat objects. One can also use here the degreewise *free* (or *projective*) objects.

Another possible generalization of condition (T) is (T″) below. It is also satisfied in the above example, but it is more difficult to obtain the desired results. (We include these considerations as a non-trivial application of left derived functors.)

(T″) If f, g are weak equivalences in $\mathcal{C}$ so is $f \otimes g$ provided f (or g) $\in \mathcal{C}_c$, i.e. $we \otimes we_c \subset we$, $we_c \otimes we \subset we$.

Let $\mathcal{C}$ be endowed with monoidal structure, such that (T″) is satisfied. Then $\otimes : \mathcal{C}^{\times 2} \to \mathcal{C}$ has a left derived $\bar{\otimes} = \widetilde{\otimes} \circ L^{\times 2} : \operatorname{Ho}\mathcal{C}^{\times 2} \to \operatorname{Ho}\mathcal{C}$, by $we_c \otimes we_c \subset we$ and (1.44). Here $L : \operatorname{Ho}\mathcal{C} \to \operatorname{Ho}\mathcal{C}_c$ is a choice-of-model functor; cf. (1.28).

Theorem (1.50)

$\operatorname{Ho}\mathcal{C}$ with $\bar{\otimes}$, K and certain canonical isomorphisms $\bar{\alpha}, \bar{\lambda}, \bar{\rho}, \bar{\tau}$ is a monoidal category.

Proof (sketch). By (T″) there are canonical isomorphisms ${}^{\mathsf{L}}\otimes \cong \bar{\otimes} \cong \otimes^{\mathsf{L}}$ where ${}^{\mathsf{L}}\otimes = \widetilde{\otimes}(L \times \mathrm{Id}) : \operatorname{Ho}\mathcal{C}^{\times 2} \xrightarrow{\sim} \operatorname{Ho}\mathcal{C}_c \times \operatorname{Ho}\mathcal{C} \to \operatorname{Ho}\mathcal{C}$. Likewise $\otimes^{\mathsf{L}} = \widetilde{\otimes}(\mathrm{Id} \times L)$. It follows that $\bar{\otimes}(\bar{\otimes} \times \mathrm{Id})$ is a left derived of $\otimes(\otimes \times \mathrm{Id})$, and analogously for $\bar{\otimes}(\mathrm{Id} \times \bar{\otimes})$, $K\bar{\otimes}$ and $\bar{\otimes}K$. Hence, using the universal property of left derived functors (1.29), we may define the "certain" natural isomorphisms to be induced by the corresponding natural isomorphisms in $\mathcal{C}$. Then it is straightforward to prove these coherent. □

Thus we are again in the situation of N° 1. In case all objects are fibrant, it is still true that $q : \boldsymbol{cComh}\,\mathcal{C} \to \mathbf{cComh}\,\mathcal{C}$ is an equivalence, cf. N° 2.

Note that here a homotopy diagonal (for any $A \in \mathcal{C}$) is a morphism $A \to A\bar{\otimes}A = LA \otimes LA$ in $\mathrm{Ho}\,\mathcal{C}$.

Appendix A Examples of Homotopical Categories

In N° 1 and 2 we recall two familiar axiomatizations of homotopy theory: Quillen's *model categories* and Baues' *cofibration categories*. We begin with the weaker axioms of Baues and show that a cofibration category is canonically a left homotopical category. Hence a model category is canonically a homotopical category. In N° 3 we discuss briefly the standard resp. rational homotopy structure of $\mathcal{S}$. In N° 4 we give conditions on a category $\mathcal{K}$ which imply that the category of simplicial objects over $\mathcal{K}$ has canonical homotopy structure. We also discuss the full subcategory of r-reduced objects, and mention some fundamental results of simplicial homotopy theory.

Diagrams are commutative unless otherwise stated.

N° 1 Cofibration categories

Let $\mathcal{C}$ be a category with given subclasses we, $cof \subset \mathrm{mor}\,\mathcal{C}$ and $\mathcal{C}_f \subset \mathrm{Ob}\,\mathcal{C}$. We assume $\mathcal{C}$ has an initial object $\emptyset$. Let $\mathcal{C}_c$ be the subclass of objects B such that $(\emptyset \to B) \in cof$. A member of cof (resp. we, resp. $cof \cap we$) is called a *cofibration* (resp. *weak equivalence*, resp. *trivial cofibration*). An object of $\mathcal{C}_f$ (resp. $\mathcal{C}_c$) is said to be *fibrant* (resp. *cofibrant*).

Definition (A.1)

We call $(\mathcal{C}, we, cof, \mathcal{C}_f)$ a *cofibration structure* in $\mathcal{C}$ if the axioms (C-1), (C-2), (C-3), (C-4) below are satisfied. A category with a distinguished cofibration structure is called a *cofibration category*.

(C-1) If f, g are maps such that gf is defined, then if two of f, g and gf are weak equivalences, so is the third. Isomorphisms are trivial cofibrations.
Cofibrations are preserved under composition.

(C-2) If $i : B \rightarrowtail A$ is a cofibration and $f : B \to X$ any map, the push-out of i along f exists in $\mathcal{C}$.
Cofibrations (resp. trivial cofibrations) are preserved under push-out.

(C-3) For every map f there is a factorization $f = p\,i$ where i is a cofibration and p a weak equivalence.

(C-4) (i) For every object X there is a fibrant model $r : X \xrightarrow{\sim} RX$. One can choose r to be a trivial cofibration.
(ii) A trivial cofibration $Y \overset{\sim}{\rightarrowtail} Z$ has a retraction if Y is fibrant.

Remark (A.2)
A cofibration category $\mathcal{C}$ is said to be *proper* if, in addition to axiom (C-2), push-out along any cofibration preserves weak equivalences. In [B] this additional property is part of the axioms, but this is unfortunate since even closed model categories need not satisfy it. One can show that $\mathcal{C}_c$ (which inherits cofibration structure in the obvious way) is always proper, cf. [B|I.1.4], [Maj|I.6.5]. In [Maj] the fundamentals of cofibration categories are developed using the axioms above. In [B] $\mathcal{C}_f$ is called a sufficiently large class of fibrant models.

Cylinder objects (A.3)
If $\mathcal{C}$ is a cofibration category, a *cylinder object* for $B \in \mathcal{C}_c$ is a left homotopy object (IB, i_0, i_1, q) such that $(i_0, i_1) : B \sqcup B \to IB$ is a cofibration. We can construct a cylinder object for any cofibrant B by factorizing the map $(1,1) : B \sqcup B \to B$ according to (C-3).
Of course $\sqcup$ denotes the categorical sum (= pushout under the initial object).

STRICT - EXTENSION LEMMA (A.4)
Let $i : A \overset{\sim}{\rightarrowtail} B$ be a trivial cofibration, and $f : A \to Y$ any map with Y fibrant. Then there is a map $g : B \to Y$ such that $g\,i = f$.
Moreover, g is unique up to homotopy rel A. □

The first part — only this is needed here — is an immediate consequence of (C-2) and (C-4) (ii). Homotopy rel A in $\mathcal{C}$ may simply be defined as homotopy in $\mathcal{C}^A$. This category of *objects under* A inherits cofibration structure in the obvious way [B]. The second part of (A.4) is an immediate consequence of the extension lemma (1.19) in $\mathcal{C}^A$.
The following is essentially known, except for the somewhat surprising part (c).

THEOREM (A.5)
Let $\mathcal{C}$ be a cofibration category, and $B \in \mathcal{C}_c$.
(a) *Two maps $f, g \in \mathcal{C}(B, X)$ are left homotopic iff there is a cylinder object IB and a homotopy $H : IB \to X$ from f to g.*
(b) *The relation $(\overset{\ell}{\sim})$ is an equivalence relation on $\mathcal{C}(B, X)$.*
(c) *Let $f, g \in \mathcal{C}(B, Y)$ with $Y \in \mathcal{C}_f$, then $f \overset{\ell}{\sim} g \Leftrightarrow f \overset{r}{\sim} g$.*
(d) *Let IB be a fixed cylinder object for B. Whenever $f, g \in \mathcal{C}(B, Y)$ with $Y \in \mathcal{C}_f$ are left homotopic, there is a homotopy $H : IB \to Y$ from f to g.*

Proof. (a) follows from (C-3). (b) is trivial except for the transitivity which follows from the fact that cylinder objects can be glued together using (C-2), cf. [Q2|p. I.1.8].
Ad (c) ($\Leftarrow$). Let $G : B \to \tilde{Y}$ be a right homotopy. Choose a cylinder object IB as in (A.3). Form the solid diagram:

$$\begin{array}{ccc} B \sqcup B & \xrightarrow{(sf,G)} & \tilde{Y} \\ {\scriptstyle (i_0,i_1)}\big\downarrow & \overset{J}{\nearrow} & \big\downarrow{\scriptstyle \sim\, p_0} \\ IB & \xrightarrow[fq]{} & Y \end{array}$$

Apply the lifting lemma in $\mathcal{C}^{B \sqcup B}$ to obtain the map J making the upper triangle commute (and the lower up to homotopy rel $B \sqcup B$). Now $H = p_1 J$ is the desired left homotopy. To prove ($\Rightarrow$) we use (a) and choose a cylinder object IB with $H : IB \to Y$, $H(i_0, i_1) = (f, g)$. We construct a right homotopy object $\tilde{Y}$ by taking the push-out

$$\begin{array}{ccc} B & \xrightarrow{f} & Y \\ {\scriptstyle i_0}\big\downarrow{\scriptstyle \sim} & \text{push} & {\scriptstyle \sim}\big\downarrow{\scriptstyle s} \\ IB & \xrightarrow[K]{} & \tilde{Y} \end{array}$$

and defining $p_0, p_1 : \tilde{Y} \rightrightarrows Y$ by $p_0 = (fq, 1_Y)$, $p_1 = (H, 1_Y)$. Note that s (the "cobase-change" of i_0) is a trivial cofibration by (C-2). Now $G = Ki_1$ is the desired right homotopy. The careful reader might argue that $\tilde{Y}$ need not be fibrant, as demanded by our definition of right homotopy object. In this case we choose a fibrant model $r : \tilde{Y} \overset{\sim}{\rightarrowtail} R\tilde{Y}$, then apply (A.4) to give $R\tilde{Y}$ a structure of right homotopy object. Now rG is the desired right homotopy. — Finally (d) follows easily from the proof of (c). For another argument see [B | II.2.2]. □

Theorem (A.6)

Let $\mathcal{C}$ be a cofibration category. Then $\mathcal{C} = (\mathcal{C}, we, \mathcal{C}_c, \mathcal{C}_f)$ is a left homotopical category.

In fact, this structure even satisfies (H-1), (H-2), (H-3) *and* (LH-4$^+$).

(LH-4$^+$) Every cofibrant object has a cylinder object.
Every cofibrant - fibrant object has a fibrant cylinder object.

Note that cylinder objects are cofibrant by definition. By (A.5) (d) cylinder objects in a cofibration category are also cylinder objects in the sense of (1.4).

Proof. Axiom (H-1) is contained in (C-1), and (H-2) follows from (C-3) and (C-4) (i). For (LH-3) we use (A.7) below; see [B | II.3] or [Maj] for a proof that a cofibration category has the properties (I), (II), (III). Axiom (LH-4) follows from (A.5) (d) which says that a cylinder object in a cofibration category is also a cylinder object in the sense of homotopical categories. Axiom (LH-4$^+$) follows from (C-4), (C-2). Finally, (A.5) (c) shows that also the "right part" of (H-3) is satisfied. □

Lemma (A.7)

If $(\mathcal{C}, we, \mathcal{C}_c, \mathcal{C}_f)$ satisfies (H-1), (H-2) *and* (I), (II), (III) *below, then there is a derived category, i.e.* (LH-3) *holds.* □

(I) For an appropriate choice of fibrant models in $\mathcal{C}$, there is for every map $f : X \to Y$ a map $Rf : RX \to RY$ with $Rf \circ r_X = r_Y \circ f$.

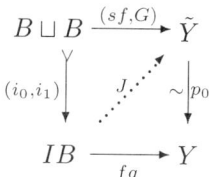

(II) For $B \in \mathcal{C}_c$, $Y \in \mathcal{C}_f$ the relation $(\overset{\ell}{\sim})$ is a natural equivalence relation on $\mathcal{C}(B, Y)$.

(III) The lifting lemma and the extension lemma hold.

Note that (II)–(III) are also necessary conditions, cf. (1.3), (1.18), (1.19). Condition (I) is used in the proof that the fairly obvious candidate for $\gamma : \mathcal{C} \to \mathrm{Ho}\,\mathcal{C}$ *is* a functor. For its construction one may use (1.25) and proof of (1.23) as a guide. For details see [B], [Q2], [Maj].

PROPOSITION (A.8)

Let $\mathcal{C}$ be a cofibration category, or any left homotopical category satisfying (LH-4$^+$). *Then $\mathcal{C}_c$, $\mathcal{C}_f$ and $\mathcal{C}_{cf}$ are left homotopical subcategories.* □

This is an extension of (1.11) and proved the same way. It implies that $\mathcal{C}_{cf}/\simeq$ is the localization of $(\mathcal{C}_{cf}, we_{cf})$. However, $\mathcal{C}_f$ and $\mathcal{C}_{cf}$ do not inherit cofibration structure in general.

The dual of (LH-4$^+$) is an axiom (RH-4$^+$), and both together give an axiom (H-4$^+$). The latter is satisfied, for instance, in any model category, see below. It implies that $\mathcal{C}_c$, $\mathcal{C}_f$ and $\mathcal{C}_{cf}$ are homotopical subcategories.

A cofibration category $\mathcal{C}$ has a rich *homotopy theory*, viewed (according to Quillen) as extra structure in $\mathrm{Ho}\,\mathcal{C}$, which comes by performing constructions in $\mathcal{C}$. Important instances of such extra structure are the (essentially well-defined) suspension functor and the family of cofiber sequences. This structure is preserved by (the derived functor of) any left homotopical functor preserving cofibrations and push-outs or, more generally, preserving homotopy push-outs [B]. The interested reader is referred to [B],[Q2],[Maj].

A dual story may be told about fibration categories.

N° 2 Model categories

We turn now to a discussion of model categories. We slightly modify Quillen's axioms [Q2]. We replace the assumption that trivial cofibrations (resp. fibrations) are preserved under push-out (resp. pull-back) by the stronger(!) assumption that a retract of a weak equivalence is a weak equivalence. Since the latter is practically always satisfied for trivial reasons, we save repeated arguments in the verification of the axioms. Moreover, using these axioms any model category has a canonically associated *closed* model structure.

Let $\mathcal{C}$ be a category with initial object $\emptyset$ and terminal object e. Let $we, cof, fib \subset \mathrm{mor}\,\mathcal{C}$. We say $X \in \mathcal{C}$ is *cofibrant* if $(\emptyset \to X) \in cof$, X is *fibrant* if $(X \to e) \in fib$.

A member of cof (resp. $cof \cap we$) is called a *cofibration* (resp. *trivial cofibration*). A member of fib (resp. $fib \cap we$) is called a *fibration* (resp. *trivial fibration*).

Definition (A.9)

We call $(\mathcal{C}, we, cof, fib)$ a *model structure* in $\mathcal{C}$ if the following axioms (M-0)–(M-5) are satisfied. A *model category* is a category with a distinguished model structure.

Let p, i be maps in $\mathcal{C}$. We say that p has the *right lifting property* (*RLP*) for i if in any solid diagram

$$(*) \qquad \begin{array}{ccc} A & \longrightarrow & C \\ {\scriptstyle i}\downarrow & \nearrow & \downarrow{\scriptstyle p} \\ B & \longrightarrow & D \end{array}$$

the dotted arrow exists. In this case we also say that i has the *left lifting property* (*LLP*) with respect to p.

(M-0) All finite limits and colimits exist (in $\mathcal{C}$).

(M-1) A cofibration has the LLP for all trivial fibrations.
A fibration has the RLP for all trivial cofibrations.

(M-2) For every map f there are factorizations
(i) $f = p\,i$ where i is a cofibration and p is a trivial fibration;
(ii) $f = q\,j$ where j is a trivial cofibration and q is a fibration.

(M-3) Cofibrations are preserved under composition and push-out.
Isomorphisms are cofibrations.
Fibrations are preserved under composition and pull-back.
Isomorphisms are fibrations.

(M-4) If f is a retract of g and g is a weak equivalence, so is f.

(M-5) If f, g are maps such that gf is defined then if two of f, g and gf are weak equivalences, so is the third.
Isomorphisms are weak equivalences.

Definition (A.10)
A model category $\mathcal{C}$ is said to be *closed* if the additional axiom (CM) is satisfied.

(CM) If f is a retract of g and g is a cofibration resp. fibration, so is f.

LEMMA (A.11)
(a) *Let $\mathcal{C}$ be a closed model category. Then the cofibrations (resp. trivial cofibrations) are precisely those maps having the LLP for all trivial fibrations (resp. fibrations). Dually, the fibrations (resp. trivial fibrations) are precisely those maps having the RLP for all trivial cofibrations (resp. cofibrations).*
(b) *For $\mathcal{C}$ to be a closed model category, it suffices that all axioms except for* (M-3) *are satisfied.*

Proof. The proof of (a) is straightforward. As it makes no use of (M-3) which is implied by (a), we see that (b) holds ([Q | p. 234]). □

PROPOSITION (A.12)
If $(\mathcal{C}, we, cof, fib)$ is a model structure, then $(\mathcal{C}, we, cof', fib')$ is a closed model structure, where cof' (resp. fib') denotes the class of all retracts of maps in cof (resp. fib).

Proof. By (A.11) (b) the only non-trivial axiom to prove for $(\mathcal{C}, we, cof', fib')$ is (M-1), but this is clear by the following observation: let f be a retract of g, then if g has the LLP (resp. RLP) for some map p, so has f. □

Remarks (A.13)

(a) The argument in the preceding proof can be used to reduce the verification of the second part of (M-1). In fact, having the first part of (M-1) and (M-2) (ii), it suffices to show that a fibration has the LLP for all maps j.

(b) One could weaken (M-0) by requiring that the push-out of a cofibration along any map and the pull-back of a fibration along any map exists. Only the formulation of the proposition above then becomes inconvenient.

THEOREM (A.14)

Let $\mathcal{C}$ be a model category. Then

(a) *Trivial cofibrations are preserved under push-out. Trivial fibrations are preseved under pull-back.*

(b) *A map f is a weak equivalence iff $\gamma(f)$ is an isomorphism.*

Proof. *Ad* (a). This is clear for closed model categories by (A.11) (a). In particular, this holds for the "closure" (A.12) of the given model structure. Hence a push-out of a trivial cofibration in $\mathcal{C}$ is a weak equivalence. By (M-3) it is also a cofibration yielding the result. — *Ad* (b). For closed model categories this is proved in [Q2|I.5]. Again the assertion follows immediately using (A.12), since Ho$\mathcal{C}, \gamma$ does only depend on *we*. □

COROLLARY (A.15)

A model category is canonically a cofibration- and a fibration category, hence a homotopical category. □

As an illustration we prove a useful criterion for the hypothesis in the Adjoint Functor Theorem (1.32).

PROPOSITION (A.16)

Let $F : \mathcal{C} \rightleftarrows \mathcal{D} : G$ be adjoint functors between model categories where $\mathcal{C}$ is closed. Suppose F satisfies: $F(cof) \subset cof, F(we) \subset we$. Then G satisfies: $G(fib) \subset fib, G(we_f) \subset we$. In particular (F, G) is an adjoint homotopical pair.

Proof. Recall (A.11) (b) that in $\mathcal{C}$ the (trivial) fibrations are characterized by a certain right lifting property. Using this one easily verifies that G preserves (trivial) fibrations. Using mapping path objects one sees that any weak equivalence in $\mathcal{D}_f$ is the composite of a right inverse of a trivial fibration followed by a trivial fibration. The second assertion follows immediately. □

N° 3 Spaces

We briefly recall the standard and the rational homotopy structure of the category $\mathcal{S}$ of spaces = simplicial sets. We call a map an *H-equivalence* if it induces an isomorphism via the functor H.

The standard homotopy structure of $\mathcal{S}$ (A.17)

It is well-known that $\mathcal{S}$ is a closed model category with the *standard* structure: *we* = weak homotopy equivalences, *cof* = injective maps, *fib* = Kan fibrations ([Q2]). In particular, $\mathcal{S}$ has a standard homotopy structure, where all objects are *cofibrant*, while the *fibrant* objects are the Kan spaces. We recall that $f : X \to Y$ is a *weak homotopy equivalence* if $f_* : \pi_*(X, x_0) \cong \pi_*(Y, f(x_0))$ for any base point

$x_0 \in X^0$. Here we define $\pi_*(X, x_0) = \pi_*(|X|, |x_0|)$ where $|X|$ is the *geometric realization* of X. There is a canonical choice of cylinder objects and of path objects in $\mathcal{S}$, cf. N° 4.

The standard homotopy structure of $\mathcal{S}_r$ (A.18)

Let $\mathcal{S}_0 = \mathcal{S}^*$ be the category of *pointed spaces* (= spaces under $*$ = simplicial pointed sets) where $* = \Delta[0]$. For general and obvious reasons ([Q 2], [B]) $\mathcal{S}_0$ inherits (closed) model structure from $\mathcal{S}$.

Let $\mathcal{S}_r \subset \mathcal{S}_0$ be the full subcategory of *r-reduced spaces* ($X^{r-1} = *$). Note that here (if $r \geqslant 1$) a map is a weak homotopy equivalence iff it is a π_*-equivalence. As is well-known, there is an Eilenberg functor (1.14) $E_r : \mathcal{S}_0 \to \mathcal{S}_r$. Hence (1.17) implies that $\mathcal{S}_r \subset \mathcal{S}_0$ is a homotopical subcategory. It is proved in [Q] that $\mathcal{S}_r$ is a closed model category, but it is fairly complicated to describe the class of fibrations explicitly.

The rational homotopy structure of $\mathcal{S}_2$ (A.19)

... is defined as follows: *we* $= \pi_* \otimes \mathbb{Q}$-equivalences, all objects are *cofibrant*, and the *fibrant* objects are the rational Kan spaces. Here a Kan space $X \in \mathcal{S}_2$ is *rational* if $\pi_*(X)$ is uniquely divisible (i.e. a $\mathbb{Q}$-module). There are several methods to show that $\mathcal{S}_2$ with the rational structure is indeed a homotopical category:

(i) The most elementary but certainly not the most efficient method is to show this directly using (A.7).

(ii) Show that this rational homotopy structure is part of a cofibration structure in $\mathcal{S}_2$ where *cof*= injective maps. The proof is sufficiently elementary. There is an analogous result for the (homotopically equivalent) category of 1-connected pointed spaces $\mathcal{S}_{(2)}$. See [Maj].

(iii) With some more effort one can show that this cofibration structure underlies a closed model structure in $\mathcal{S}_2$. The class of fibrations contains all one would expect, but also strange things like injective fibrations; see [Q].

(iv) The category $\mathcal{S}$ has a convenient closed model structure with *we* $= H_*(-;\mathbb{Q})$-equivalences and *cof*= injective maps (see [Bou] where an analogous result is proved for any homology theory). Thus $\mathcal{S}_2$ may be regarded as a homotopical subcategory of $\mathcal{S}$ (with respect to rational homotopy structure).

N° 4 Simplicial objects

Given a category $\mathcal{K}$, we give conditions for the existence of a standard homotopy structure in ${}^{\Delta}\mathcal{K}$. We give here essentially an account of Quillen's work on the same subject, but we simplify the matters in that we do not discuss *simplicial categories* in general. Moreover we assume the existence of *one* projective generator in $\mathcal{K}$, while Quillen obtains similar results when there is *a set* of projective generators.

Examples are given in (A.27).

Let $\mathcal{K}$ be a category with (co)limits and a projective generator.

Here $R \in \mathcal{K}$ is a *generator* if every object is the target of an effective epimorphism with source a coproduct of copies of R. An object is *projective* if every effective epimorphism to this object has a section. An *effective epimorphism* is a map $f : X \to Y$ such that $f^* : \mathrm{mor}(Y, U) \to \mathrm{mor}(X, U)$ is injective (all $U \in \mathcal{K}$) and $\mathrm{Im} f^*$ is the set of all maps $g : X \to U$ being *coarser* than f (i.e. $f\alpha = f\beta \Rightarrow g\alpha = g\beta$).

An object $A \in \mathcal{K}$ is said to be *free* if A is an arbitrary coproduct of copies of R. This notion depends on the choice of the projective generator R, but in all examples of interest there is a canonical choice. In any case the retracts of all free objects are precisely the projective objects.

Exterior product, exterior function complex, Hom-complex (A.20)

Let X be a space and A a simplicial object over $\mathcal{K}$. The formula $(X \otimes A)_n = \bigsqcup_{\sigma \in X_n} A_n$ defines a functor

$$\otimes : \mathcal{S} \times {}^{\Delta}\mathcal{K} \longrightarrow {}^{\Delta}\mathcal{K}.$$

Now $X \otimes -$ has a right adjoint $-^X$ which clearly yields a functor

$$-^- : \mathcal{S}^{\mathrm{op}} \times {}^{\Delta}\mathcal{K} \longrightarrow {}^{\Delta}\mathcal{K}.$$

Also $- \otimes A$ has a right adjoint $\mathrm{Hom}(A, -)$ which yields a functor

$$\mathrm{Hom}(-,-) : ({}^{\Delta}\mathcal{K})^{\mathrm{op}} \times {}^{\Delta}\mathcal{K} \longrightarrow \mathcal{S}$$

cf. [Q2|II.1], [K2]. In case ${}^{\Delta}\mathcal{K} = \mathcal{S}$ it is easily verified that $\otimes$ is the ordinary product, and that $Y^X = \mathrm{Hom}(X, Y)$ is the ordinary function complex of spaces.

Definition (A.21)

We call $f : B \to A \in {}^{\Delta}\mathcal{K}$ a *free map* if (for all $n \geqslant 0$) $A_n = B_n \sqcup \bigsqcup_{\nu \in M_n} R$, via f_n and some (possibly empty) set M_n of maps $R \to A_n$, such that:

(D) $\nu \in M_n \Rightarrow s_i\nu \in M_{n+1}$ $(i = 0, \dots, n)$.

An object A in ${}^{\Delta}\mathcal{K}$ is *free* if $\emptyset \to A$ is a free map.

One may show that f is free iff it is a colimit of maps $B \to A^n$ where A^n is obtained by "attaching" copies of $\Delta[n] \otimes R$ to A^{n-1} via maps $\dot{\Delta}[n] \otimes R \to A^{n-1}$. Here R is regarded as a constant simplicial object; cf. [Q2|p. II.4.11], [K2].

Regarding R as a constant simplicial object, $\mathrm{Hom}(R, -) : {}^{\Delta}\mathcal{K} \to \mathcal{S}$ is just the prolongation (= degreewise application) of $\mathcal{K}(R, -)$. Consider the following conditions that $\mathcal{K}$ with R may satisfy or not:

$(*)$ $\mathrm{Hom}(R, A)$ is a Kan complex for all $A \in {}^{\Delta}\mathcal{K}$.

$(**)$ $\mathcal{K}(R, -)$ preserves filtered colimits.

Note that $\mathrm{Hom}(R, -)$ has the left adjoint $- \otimes R$ which is simply the functor which associates with X the free simplicial object generated by X,

(A.22) $$- \otimes R : \mathcal{S} \rightleftarrows {}^{\Delta}\mathcal{K} : \mathrm{Hom}(R, -).$$

In $\mathcal{S}$ we consider the *standard* model structure (A.17). (We do not know whether the following holds for other structures as well.)

THEOREM (A.23)

Let one of the conditions $(*)$ *or* $(**)$ *be satisfied. Then* ${}^{\Delta}\mathcal{K}$ *is a model category in the following obvious way:*

we = maps f for which $\mathrm{Hom}(R, f)$ is a weak equivalence in $\mathcal{S}$

fib = maps f for which $\mathrm{Hom}(R, f)$ is a fibration in $\mathcal{S}$

cof = free maps.

Proof. Quillen proves that endowed with the "closure" of this structure (the cofibrations are the *retracts* of the free maps (A.12)) ${}^{\Delta}\mathcal{K}$ is a closed model category, see [Q2|II.4, Thm. 4]. From this the assertion follows, since clearly free maps are preserved under composition and push-out. □

In fact ${}^{\Delta}\mathcal{K}$ is a closed *simplicial* model category, i.e. two more axioms are satisfied concerning the functors in (A.20). The precise meaning of this is not needed, but the following consequence ([Q2|p. II.2.6]).

ADDENDUM (A.24)

$\Delta[1] \otimes B$ is a cylinder object and $C^{\Delta[1]}$ a path object, for any cofibrant B and fibrant C in ${}^{\Delta}\mathcal{K}$. Hence: *The homotopy relation in ${}^{\Delta}\mathcal{K}$ is given by simplicial homotopy.* □

Here $i_0, i_1 : B \rightrightarrows \Delta[1] \otimes B$ are of course induced by $(0), (1) : \Delta[0] \rightrightarrows \Delta[1]$ via $- \otimes B$, etc. A map $H : \Delta[1] \otimes B \to A$ is called a *simplicial homotopy* from Hi_0 to Hi_1 in ${}^{\Delta}\mathcal{K}$.

Let in addition $\mathcal{K}$ be pointed: initial object = terminal object = $*$.

Let $r \geqslant 1$. We denote by $({}^{\Delta}\mathcal{K})_r$ the category of r-reduced simplicial objects over $\mathcal{K}$. Here X is *r-reduced* if the unique map $* \to X$ is isomorphic in degrees $< r$.

THEOREM (A.25)

The category $({}^{\Delta}\mathcal{K})_r \subset {}^{\Delta}\mathcal{K}$ is a homotopical subcategory, in which $\Delta[1] \otimes B$ is a cylinder object and $E_r(C^{\Delta[1]})$ is a path object. □

Proof. This is an application of (1.17) using the following facts, which are proved in [Q|p. 153 ff].

(i) In the factorization of any map $f : B \to C \in {}^{\Delta}\mathcal{K}$ into a free map $B \to A$ followed by a trivial fibration $A \to C$, one may assume A to be r-reduced if C is so.

(ii) There is a functor $E_r : {}^{\Delta}\mathcal{K} \to ({}^{\Delta}\mathcal{K})_r$ which is left inverse - right adjoint to the embedding. There is a natural isomorphism $\mathrm{Hom}(A, E_r C) = E_r \mathrm{Hom}(A, C)$.

The last formula indicates a method of construction. It follows immediately from (ii) (put $A = R$) that if C is a fibrant object, so is $E_r C$ and $E_r C \to C$ is a weak equivalence if $\mathrm{Hom}(R, C)$ is $(r-1)$-connected. Similarly, E_r preserves weak equivalences between fibrant objects. Hence E_r is an Eilenberg functor (1.14).

(iii) The canonical cylinder object $\Delta[1] \otimes B$ in $\mathcal{K}$ is r-reduced if B is so. □

Remark (A.26)

One may also deduce that $({}^{\Delta}\mathcal{K})_r$ is a cofibration category where *cof* = free maps, and $\Delta[1] \otimes B$ is a cylinder object in the sense of cofibration categories.— Quillen shows more in case $(*)$: $({}^{\Delta}\mathcal{K})_r$ is a model category where *fib* = maps f for which $\mathrm{Hom}(R, f)$ is a fibration in $\mathcal{S}_r$ (or equivalently, f has the RLP for all trivial cofibrations). Moreover, $E_r(C^{\Delta[1]})$ is a path object in the sense of model categories.

Examples (A.27)

(a) For $\mathcal{K} = \langle \mathrm{Set} \rangle$ resp. $\langle \mathrm{Set} \rangle^*$ condition $(**)$ holds with $R = *$ resp. $* \sqcup *$. We re-obtain the standard model structure of $\mathcal{S}$ resp. $\mathcal{S}_0$. Furthermore $\mathcal{S}_r$ is a homotopical subcategory.

(b) Let $\mathcal{K} = \langle \mathrm{Gp} \rangle$ resp. $\langle \mathrm{LA} \rangle$. It is clear that limits and colimits exist, and that $R = \mathbb{Z}$ resp. K is a projective generator. Here condition $(*)$ holds because a sGp is a Kan space (i.e. fibrant in standard $\mathcal{S}$). Notice that a map in $\langle \mathrm{sGp} \rangle$ is a fibration iff it is Kan fibration in $\mathcal{S}$, thus (A.28) gives another characterization. Similarly for $\langle \mathrm{sLA} \rangle$.

(c) The category $\mathcal{K} = \langle \mathrm{CHA} \rangle$ of rigid normal complete Hopf algebras has arbitrary limits and colimits, and the projective generator $R = \mathsf{K}\langle\langle \mathbf{X} \rangle\rangle$, cf. (3.70). The natural bijection $\mathrm{CHA}(\mathsf{K}\langle\langle \mathbf{X} \rangle\rangle, A) \approx \mathcal{P}A$ (3.46) implies $(*)$. We also see that a map f in $\langle \mathrm{sCHA} \rangle$ is a weak equivalence iff $\mathcal{P}_{.} f$ is a π_*-equivalence, and a fibration iff $N(\mathcal{P}_{.} f)$ is surjective in degrees > 0. Here one can also use the functor $\mathcal{G}$ instead of $\mathcal{P}$.

The following is proved in [Q2|p. II.3.8].

PROPOSITION (A.28)
A map $f \in \langle \mathrm{sGp} \rangle$ is a Kan fibration iff $N(f)$ is surjective in degrees > 0. □

The *normalization* of a sGp G is a dg group NG (the boundaries form a normal subgroup of the cycles), defined as in the abelian case (B.5). Recall that the homotopy groups of G, considered as a (Kan) space, are given by $\pi_*(G) = H_*(NG)$. In Chp. 4 this fact is of fundamental importance, in particular in the context of simplicial modules.

We also need the following two famous results of simplicial homotopy theory.

LEMMA (A.29)
Let $F : \mathcal{K} \to \mathcal{M}$ be any functor between categories with finite coproducts. Then the prolongation $F_{.} : {}^{\Delta}\mathcal{K} \to {}^{\Delta}\mathcal{M}$ preserves simplicial homotopy.

More precisely, if $H : \Delta[1] \otimes B \to C$ is a simplicial homotopy from f to g in ${}^{\Delta}\mathcal{K}$, there is a canonical simplicial homotopy $\tilde{H} : \Delta[1] \otimes F_{.}B \to F_{.}C$ from $F_{.}f$ to $F_{.}g$. □

One only has to observe that there is a canonical map $X \otimes (F_{.}B) \to F_{.}(X \otimes B)$ in ${}^{\Delta}\mathcal{M}$. In general this map is not an isomorphism, cf. [Q2|p. II.1.11].

The second is a deeper result: the *connectivity theorem* of Curtis ([Cu|4.10]). Here we only need that $\mathcal{K}$ is a pointed category with coproducts, R an object of $\mathcal{K}$. Then (R-) free objects are defined in ${}^{\Delta}\mathcal{K}$ and we have a *Milnor construction* $F_{.} : \mathcal{S}_0 \to {}^{\Delta}\mathcal{K}$, $F_{.}X = \mathrm{coker}(* \otimes R \hookrightarrow X \otimes R)$. Let $\mathcal{V}$ be an abelian category.

THEOREM (A.30)
Let $\Phi : \mathcal{K} \to \mathcal{V}$ be a functor which preserves filtered colimits and initial object. Suppose that $\Phi_{.}F_{.}X$ is n-connected when X is a finite bouquet of 1-spheres. Then $\Phi_{.}A$ is $(r+n-1)$-connected for every r-reduced free object A in ${}^{\Delta}\mathcal{K}$. □

Chapter 2

Differential Algebra

In this chapter $\mathcal{V}$ is an additive monoidal category, i.e. it is additive and there is given a "tensor product", i.e. a bi-additive functor $\otimes : \mathcal{V} \times \mathcal{V} \to \mathcal{V}$ which is symmetric, associative and unital with some object $K \in \mathcal{V}$. Unless otherwise stated, we assume that $\mathcal{V}$ is strongly additive, i.e. arbitrary direct sums exist and are preserved by $\otimes$ in each variable. This assumption is not needed if all (co)chain complexes under consideration are bounded below. Occasionally we need other colimits or limits in $\mathcal{V}$.

For precise definitions and explanations, see § 1, N° 1 and App. B, N° 1. Of course, our main interest is the case $\mathcal{V} = \mathcal{M}od_{\mathsf{K}}$. But the much greater generality does not cause extra complications and is useful, for instance, in connection with questions of duality. The latter is investigated in § 3, N° 3.

§ 1 PRELIMINARIES

N° 1 Chain complexes

Let $\mathcal{V}$ be an *additive* category, i.e. on each set $\mathcal{V}(V, W)$ there is defined an abelian group structure, such that composition is bi-linear; moreover finite coproducts exist. In particular, a *zero object* (i.e. initial object) exists, it is denoted by 0. One checks that this is also a terminal object. We write $\oplus$ for *direct sums* = coproducts in $\mathcal{V}$. One checks that finite direct sums are also finite products in the obvious way, and conversely. For more details see [Mc].

Chain complexes (2.1)

A *graded object* is a family $\{X_n\}_{n\in\mathbb{Z}}$ where $X_n \in \mathcal{V}$. A *degree k map* $f : X \to Y$ between graded objects is a family $\{f_n : X_n \to Y_{n+k}\}_{n\in\mathbb{Z}}$ where $f_n \in \mathrm{mor}\mathcal{V}$. Let $\mathrm{Hom}(X, Y)$ be the graded abelian group with $\mathrm{Hom}_k(X, Y) = \{$degree k maps $X \to Y\}$. Any $f \in \mathrm{Hom}(X, Y)$ is called a *graded map*, the degree of f is denoted by $|f|$. Clearly $|g \circ f| = |f| + |g|$ for $g \in \mathrm{Hom}(Y, Z)$.

A *chain complex* or *differential graded object* over $\mathcal{V}$, is a graded object V endowed with a *differential*, i.e. a map $d \in \mathrm{Hom}_{-1}(V, V)$ satisfying $dd = 0$. With the obvious notion of chain map these objects form a category $\langle \mathrm{DG} \rangle$. The category of graded objects $\langle \mathrm{G} \rangle$ with degree 0 maps as morphisms is regarded as the full subcategory of $\langle \mathrm{DG} \rangle$ consisting of all chain complexes with differential = 0. We call $V \in \langle \mathrm{DG} \rangle$ *positive* if $V_{<0} = 0$, these form a full subcategory $\langle \mathrm{dg} \rangle \subset \langle \mathrm{DG} \rangle$.

The complex of graded maps (2.2)

For V, $W \in \langle \mathrm{DG} \rangle$ we regard $\mathrm{Hom}(V,W)$ as a chain complex over $\mathcal{M}od_{\mathbb{Z}}$ with differential $D(f) = df - (-1)^{|f|} fd$. Note that $f \in \mathrm{Hom}_0(V,W)$ is a *chain map* iff $D(f) = 0$. A map $H \in \mathrm{Hom}_1(V,W)$ with $D(H) = f - g$ is said to be a *chain homotopy* from f to g. One checks that with respect to composition of graded maps D is a derivation.

Any graded map $f : V \to W$ induces a graded map $f^* : \mathrm{Hom}(W,U) \to \mathrm{Hom}(V,U)$ defined by $f^*(\alpha) = (-1)^{|f||\alpha|}\alpha f$. Any graded map $g : U \to T$ induces a graded map $g_* : \mathrm{Hom}(V,U) \to \mathrm{Hom}(V,T)$ defined by $g_*(\beta) = g\beta$.

One checks the formulas $g_* f^* = (-1)^{|g||f|} f^* g_*$, $(ef)^* = (-1)^{|e||f|} f^* e^*$, $(hg)_* = h_* g_*$ and concludes that $D(f^*) = D(f)^*$, $D(g_*) = D(g)_*$.

Cochain complexes (2.3)

For graded objects (resp. graded maps) X we use the notation

$$X^n = X_{-n}\,.$$

A map $X^* \to Y^*$ of (upper) degree k is, by definition, the same as a map $X_* \to Y_*$ of degree $-k$. In this way $\langle \mathrm{DG} \rangle$ is identified with the category $\langle \mathrm{DG}^* \rangle$ of *cochain complexes*.

Let $\langle \mathrm{dg}^* \rangle \subset \langle \mathrm{DG}^* \rangle$ be the full subcategory of *positive* cochain complexes (= *negative* chain complexes), i.e. of all V^* with $V^{<0} = 0$.

Let $\mathcal{V}$ be a strongly additive monoidal category, see App. B, N° 1. If one works exclusively with bounded below graded objects, then additive monoidal is sufficient.

Graded tensor product (2.4)

The category $\langle \mathrm{G} \rangle$ is again an additive monoidal category as follows. First define $V \otimes W$ by $(V \otimes W)_n = \bigoplus_k V_k \otimes W_{n-k}$ and similar for maps (see below). The unit object (again denoted by K) and the natural isomorphisms $\alpha, \lambda, \rho, \tau$ are obvious. In the definition of τ the usual sign rule appears, i.e. $\tau_{V,W}$ is the direct sum of the maps $(-1)^{ij}\tau_{V_i,W_j}$. In order to be able to define α we need one of the extra assumptions (strongly additive, or bounded below).

For maps f, g of any degree $f \otimes g$ is defined by $(f \otimes g)_n = \sum_k (-1)^{|g|k} f_k \otimes g_{n-k}$ (*note the sign rule!*) which has degree $|f| + |g|$. This extended "$\otimes$" for graded maps is additive in each variable and satisfies $(d \otimes e)(f \otimes g) = (-1)^{|e||f|} df \otimes eg$. The maps α, λ, ρ are also natural with respect to graded maps, and $\tau_{X,Y}(f \otimes g) = (-1)^{|f||g|}(g \otimes f)\tau_{V,W}$ for $f : V \to X$, $g : W \to Y$.

Now also $\langle \mathrm{DG} \rangle$ is an additive monoidal category, where $V \otimes W$ is given the differential $d \otimes 1 + 1 \otimes d$. One checks the formula $D(f \otimes g) = Df \otimes g + (-1)^{|f|} f \otimes Dg$ for graded maps f, g between chain complexes. For $V \in \langle \mathrm{DG} \rangle$ define $V^{\otimes n}$ by $V^{\otimes 0} = K$, $V^{\otimes 1} = V$, $V^{\otimes n+1} = V^{\otimes n} \otimes V$ and analogously for graded maps.

Shifting (2.5)

For $V \in \langle \mathrm{G} \rangle$ we define $s_n V \in \langle \mathrm{G} \rangle$ by $(s_n V)_k = V_{k-n}$. A graded map $f : V \to W$ yields (for any n, m) the "n,m-shift" $\langle f \rangle : s_n V \to s_m W$, i.e. the graded map with k-th component f_{k-n}, and with degree $|f| - n + m$.

For $V \in \langle \mathrm{DG} \rangle$ we endow $s_n V \in \langle \mathrm{DG} \rangle$ with the differential $(-1)^n \langle d_V \rangle$. Then $s_n V \otimes s_m W = s_{n+m}(V \otimes W)$. Moreover, for an n,m-shift $\langle f \rangle$ as above we have $D\langle f \rangle = (-1)^m \langle Df \rangle$.

For objects with upper grading V^* we define $s^n V^*$ by $(s^n V^*)^k = V^{k-n}$ etc. Note that $s^1 V^*$ is identified with $s_{-1}V_*$ according to (2.3).

N° 2 DG (co)algebras

A *DG algebra* or *DGA* is a monoid over $\langle \mathrm{DG} \rangle$, see (B.1). An *augmented DGA* is a based monoid. We obtain categories $\langle \mathrm{DGA} \rangle$ resp. $\langle \mathrm{DGA} \rangle_0$. Restricting to *positive* objects we obtain full subcategories $\langle \mathrm{dgA} \rangle \subset \langle \mathrm{DGA} \rangle$ etc. Restricting to objects which are zero in degrees $< r$ we obtain the full subcategory $\langle \mathrm{dgA} \rangle_r \subset \langle \mathrm{dgA} \rangle_0$ of *r-reduced* objects ($r \geqslant 1$). Restricting to objects with differential $= 0$ we obtain the full subcategory $\langle \mathrm{GA} \rangle \subset \langle \mathrm{DGA} \rangle$ of *graded algebras*. Analogously in the augmented case.

In a dual manner, we define a *DG coalgebra* or *DGC* to be a comonoid over $\langle \mathrm{DG} \rangle$. A *coaugmented DGC* is a based comonoid. We obtain categories $\langle \mathrm{DGC} \rangle$ resp. $\langle \mathrm{DGC} \rangle_0$.

Definitions (2.6)

Let $f,\, g : A \rightrightarrows A' \in \langle \mathrm{GA} \rangle$. A graded map $\nabla : A \to A'$ is an *(f,g)-derivation* if

$$\nabla m_A = m_{A'}(\nabla \otimes g + f \otimes \nabla)\,.$$

It follows that $\nabla \eta = 0$. If we work in $\langle \mathrm{GA} \rangle_0$ then also $\varepsilon \nabla = 0$ is required. A $(1,1)$-derivation $A \to A$ is also called a *derivation* (resp. a *differential* if its square is zero and its degree is -1). Note that a DGA is just a GA endowed with a differential.

In a dual manner we define *(f,g)-coderivations* where $f,\, g \in \langle \mathrm{GC} \rangle$ etc.

LEMMA (2.7)

(a) *Let $\alpha : A' \to A''$, $\beta : \tilde{A} \to A \in \langle \mathrm{GA} \rangle$ and $\nabla : A \to A'$ be an (f,g)-derivation. Then $\alpha \nabla \beta$ is an $(\alpha f \beta, \alpha g \beta)$-derivation.*

(b) *A specialty of derivations $\partial : A \to A$ of odd degree is that $\partial\partial$ is again a derivation.*

(c) *The (f,g)-derivations form a sub-graded module of* $\mathrm{Hom}(A, A')$, *closed under the differential D (2.2) in case $A, A' \in \langle \mathrm{DGA} \rangle$.* □

Analogously for (f,g)-coderivations.

There is a canonical homotopy notion in $\langle \mathrm{DGA} \rangle$ resp. $\langle \mathrm{DGA} \rangle_0$.

Definition (2.8)

Given $f,\, g : A \rightrightarrows A' \in \langle \mathrm{DGA} \rangle$ resp. $\langle \mathrm{DGA} \rangle_0$, we call f *derivation homotopic* to g ($f \simeq g$) if there is an (f,g)-derivation $H \in \mathrm{Hom}_1(A, A')$ satisfying $D(H) = f - g$. Then H is called a *derivation homotopy* from f to g ($H : f \simeq g$).

In a dual manner we define a *coderivation homotopy* relation for maps in $\langle \mathrm{DGC} \rangle_0$.

The cup-product (2.9)

Let $C \in \langle \mathrm{GC} \rangle$ and $A \in \langle \mathrm{GA} \rangle$. Then $\mathrm{Hom}(C, A)$ is a graded algebra over $\mathcal{M}od_{\mathbb{Z}}$ with multiplication

$$v \cup w = m_A(v \otimes w)\Delta_C\,, \qquad v,\, w \in \mathrm{Hom}(C, A)$$

called the *cup-product*. The unit is $\mathbb{1} = \eta\varepsilon \in \mathrm{Hom}_0(C, A)$. The verification of this and the following naturality properties is left to the reader.

(i) $\alpha(v \cup w) = \alpha v \cup \alpha w\,,\ \alpha\mathbb{1} = \mathbb{1}\,,$

(ii) $\nabla(v \cup w) = \nabla v \cup gw + (-1)^{|\nabla||v|} fv \cup \nabla w\,,\ \nabla\mathbb{1} = 0\,,$

(i′) $(v \cup w)\gamma = v\gamma \cup w\gamma\,,\ \mathbb{1}\gamma = \mathbb{1}\,,$

(ii′) $(v \cup w)\nabla = vf \cup w\nabla + (-1)^{|\nabla||w|} v\nabla \cup wg\,,\ \mathbb{1}\nabla = 0\,,$

where $\alpha : A \to A' \in \langle \mathrm{GA} \rangle$, $\gamma : C' \to C \in \langle \mathrm{GC} \rangle$, $\nabla : A \to A'$ an (f, g)-derivation and $\nabla : C' \to C$ an (f, g)-coderivation. These formulas say that α_* and γ^* are graded algebra maps, while ∇_* resp. ∇^* is an (f_*, g_*)-derivation resp. (f^*, g^*)-derivation. It follows that for $C \in \langle \mathrm{DGC} \rangle$, $A \in \langle \mathrm{DGA} \rangle$ we have that $(\mathrm{Hom}(C, A), \cup, D)$ is a DGA over $\mathcal{M}od_{\mathbb{Z}}$, i.e. D is a derivation.

The (co)augmentation (co)ideal (2.10)

Assume that kernels exist in $\mathcal{V}$. For $A \in \langle \mathrm{DGA} \rangle_0$ we define $j : \overline{A} \to A$ by $j = \ker \varepsilon$. From $\varepsilon\eta = 1$ it follows that $A = K \oplus \overline{A}$ in a canonical way.

Dually, for $C \in \langle \mathrm{DGC} \rangle_0$ we define $q : C \to \widetilde{C}$ by $q = \operatorname{coker} \eta$. We have $C = K \oplus \widetilde{C}$. We may also regard $\widetilde{C}$ as a kernel. Let $j : \widetilde{C} \to C$ be induced by $1 - \eta\varepsilon : C \to C$, then $j = \ker \varepsilon$ and $qj = 1$. Let $\iota_1, \iota_2 : C \rightrightarrows C \otimes C$ be the chain maps $\iota_1 = (1 \otimes \eta)\rho_C^{-1}$, $\iota_2 = (\eta \otimes 1)\lambda_C^{-1}$ where $\rho_C : C \otimes K \cong C$, $\lambda_C : K \otimes C \cong C$ are canonical.

The reduced diagonal (2.11)

Now the diagonal Δ of C induces the DG-map $\widetilde{\Delta} : \widetilde{C} \to \widetilde{C} \otimes \widetilde{C}$ (the *reduced diagonal*). One verifies $(j \otimes j)\widetilde{\Delta} = (\Delta - \iota_1 - \iota_2)j$, i.e. regarding $\widetilde{C}$ as a kernel, $\widetilde{\Delta}$ is the restriction of $\Delta - \iota_1 - \iota_2$. Often this is used as a definition of the reduced diagonal.

Cocompleteness (2.12)

Assume that kernels exist in $\mathcal{V}$. For $C \in \langle \mathrm{DGC} \rangle_0$ define chain complexes $F_{n-1}C = \ker(q^{\otimes n}\Delta^{(n)})$ where $n \geqslant 1$, $\Delta^{(1)} = \mathrm{id}_C$, $\Delta^{(n)} = (\Delta^{(n-1)} \otimes \mathrm{id})\Delta : C \to C^{\otimes n}$. (Recall there is a general coassociativity law for the iterated diagonal $\Delta^{(n)}$.) Now C is *cocomplete* if this so-called *coaugmental filtration* $F_0C \subset F_1C \subset \ldots$ is cocomplete, i.e. $C = \operatorname{colim}\{F_nC\}$. Note that $F_0C = K$ and $F_1C = K \oplus \mathcal{P}C$.

Often it is more appropriate to consider the produced filtration of $\widetilde{C}$ given by $F_n\widetilde{C} = F_nC/K$ which is the kernel of $\widetilde{\Delta}^{(n)} : \widetilde{C} \to \widetilde{C}^{\otimes n}$.—For example if C is 1-reduced then $C_{<n} \subset F_{n-1}C$ and thus C is cocomplete.

Remark. We use here the notation $V \subset W$ for a *normal monomorphism* (= kernel) in an additive category, cf. (C.17). Of course, there is a dual notion of complete augmented DGA.

Connectedness (2.13)

The following variant is worth mentioning. We define C to be *c-connected* if it admits a cocomplete filtration $\{F_nC\}$ with $F_0C = K$ and $\widetilde{\Delta}|F_n\widetilde{C}$ factors through $(F_{n-1-c}\widetilde{C})^{\otimes 2}$. Notice that r-reduced implies $(r-1)$-connected. Moreover, connected implies cocomplete provided filtered colimits (in $\mathcal{V}$) preserve normal monomorphisms. The converse holds under certain flatness or splittability assumptions. For instance, in case $\mathcal{V}$ is the category of flat K-modules these notions are equivalent.

N° 3 Tensor (co)algebras

Free based monoids may be defined over any monoidal category with coproducts. For $V \in \langle \mathrm{DG} \rangle$ this is the *tensor algebra* $\mathsf{T}V = \bigoplus_{n=0}^{\infty} V^{\otimes n}$. The multiplication of $\mathsf{T}V \in \langle \mathrm{DGA} \rangle_0$ is given by the canonical isomorphisms $m^{k,n-k} : V^{\otimes k} \otimes V^{\otimes n-k} \approx V^{\otimes n}$.

Notice that cofree based comonoids must be defined using a product instead of a direct sum. We prefer here the cocomplete variant of this where again the direct sum is used. Let $i^k : V^{\otimes k} \rightleftarrows \mathsf{T}V : p^k$ be the canonical maps.

Tensor coalgebra (2.14)

Let $\mathsf{T}'V = (\mathsf{T}V, \Delta) \in \langle \mathrm{DGC} \rangle_0$ be defined as follows. The counit is $\varepsilon = p^0$, the coaugmentation $\eta = i^0$. The diagonal is given by $\Delta i^n = \Delta^n$ where

$$\Delta^n = \sum \Delta^{k,n-k} : V^{\otimes n} \to \bigoplus_{k=0}^{n} V^{\otimes k} \otimes V^{\otimes n-k}$$

with $\Delta^{k,n-k} : V^{\otimes n} \approx V^{\otimes k} \otimes V^{\otimes n-k}$ the canonical isomorphism which is inverse to $m^{k,n-k}$.

PROPOSITION (2.15)

$F_n(\mathsf{T}'V) = \bigoplus_{k=0}^{n} V^{\otimes k}$, thus: $\mathsf{T}'V$ *is cocomplete and* $\mathcal{P}(\mathsf{T}'V) = V$.

A similar result holds for $\mathsf{T}'V \otimes \mathsf{T}'W$, in particular the primitives are $V \oplus W$.

Proof. This follows easily from the fact that $q^{\otimes n}\Delta^{(n)} : \mathsf{T}V \to (\mathsf{T}V)^{\otimes n} \to (\widetilde{\mathsf{T}V})^{\otimes n}$ is the direct sum of two maps: the former is a zero-map on $\bigoplus_{k=0}^{n-1} V^{\otimes k}$, the other has a retraction and thus kernel 0. The former is the direct sum for $k < n$ of the maps

$$V^{\otimes k} \xrightarrow{\Delta^{(n)}} \bigoplus_{\sum k_i = k} V^{\otimes k_1} \otimes \ldots \otimes V^{\otimes k_n} \xrightarrow{q^{\otimes n}} \bigoplus_{\sum k_i = k, k_i > 0} V^{\otimes k_1} \otimes \ldots \otimes V^{\otimes k_n},$$

and the other is the direct sum for $k \geqslant n$ of these maps. (To be more precise, insert some parantheses and an associativity isomorphism for the tensor product.) □

The following result shows that the tensor algebra is indeed "free" in some sense. The tensor coalgebra is "cofree" as a cocomplete coalgebra. Note that V reduced implies that $\mathsf{T}'V = \prod V^{\otimes n}$ which is cofree as a coalgebra. The details are omitted.

THEOREM (2.16)

(a) *Let* $V \in \langle \mathrm{DG} \rangle$ *and* $A \in \langle \mathrm{DGA} \rangle_0$. *For any chain map* $f^1 : V \to \overline{A}$ *there is unique* $f : \mathsf{T}V \to A \in \langle \mathrm{DGA} \rangle_0$ *with* $fi^1 = f^1$.

Given $f, g : \mathsf{T}V \rightrightarrows A \in \langle \mathrm{DGA} \rangle_0$ *and graded map* $\nabla^1 : V \to \overline{A}$, *there is unique* (f, g)*-derivation* $\nabla : \mathsf{T}V \to A$ *with* $\nabla i^1 = \nabla^1$.

(b) *Let* $V \in \langle \mathrm{DG} \rangle$ *and* $C \in \langle \mathrm{DGC} \rangle_0$. *Suppose* C *is cocomplete (or* V *reduced). For any chain map* $f^1 : \widetilde{C} \to V$ *there is unique* $f : C \to \mathsf{T}'V \in \langle \mathrm{DGC} \rangle_0$ *with* $p^1 f = f^1$.

Given $f,\, g : C \rightrightarrows \mathsf{T}'V \in \langle\mathrm{DGC}\rangle_0$ *and graded map* $\nabla^1 : C \to V$ *with* $\nabla^1\eta = 0$, *there is unique* (f,g)*-coderivation* $\nabla : C \to \mathsf{T}'V$ *with* $p^1\nabla = \nabla^1$. □

Remark (2.17)

Suppose that $A \cong (\mathsf{T}V, d)$ for some V and d. Then an (f,g)-derivation H satisfies $D(H) = f - g$ iff this holds on $V \subset A$. This follows from (2.16), since both $D(H)$ and $f - g$ are (f,g)-derivations. Similarly, $\alpha : A \to A' \in \langle\mathrm{GA}\rangle_0$ satisfies $D(\alpha) = 0$ iff this holds on $V \subset A$, see (2.7) (a).

§ 2 TWISTING MAPS AND THE (CO)BAR CONSTRUCTION

N° 1 Twisting maps and homotopies

Let $C \in \langle\mathrm{DGC}\rangle_0$, $A \in \langle\mathrm{DGA}\rangle_0$. Then $v \in \mathrm{Hom}_{-1}(C, A)$ is said to be a *twisting map* if

$$D(v) = v \cup v\,, \quad v\eta = 0\,,\ \varepsilon v = 0\,.$$

The set of twisting maps $C \to A$ is denoted by $\mathrm{T}(C, A)$.

Given $v,\, w \in \mathrm{T}(C, A)$ we say that v is *homotopic* to w (and write $v \simeq w$) if there is $h \in \mathrm{Hom}_0(C, A)$ satisfying

$$D(h) = v \cup h - h \cup w\,, \quad h\eta = \eta\,,\ \varepsilon h = \varepsilon\,.$$

We call such h a *twisting homotopy* from v to w (and write $h : v \simeq w$).

Naturality: Let $v \in \mathrm{T}(C, A)$, $\alpha : A \to A' \in \langle\mathrm{DGA}\rangle_0$ and $\gamma : C' \to C \in \langle\mathrm{DGC}\rangle_0$. Then $\alpha v \gamma \in \mathrm{T}(C', A')$. Given $h : v \simeq w$ then $\alpha h \gamma : \alpha v \gamma \simeq \alpha w \gamma$.

Remark (2.18)

It is easy to see that this relation ($\simeq$) is reflexive and transitive. In fact, the cup-product defines a natural "addition" for twisting homotopies. If C is cocomplete or A complete (2.12), then ($\simeq$) is also symmetric. In fact, $h = \mathbb{1} + \bar{h}$ has the inverse twisting homotopy $\mathbb{1} - \bar{h} + \bar{h}^2 - \dots$, which is a finite sum on each F_nC by definition and thus well-defined on C.

LEMMA (2.19)

Let $t \in \mathrm{T}(C, A)$ *and* $f,\, g : A \rightrightarrows A' \in \langle\mathrm{DGA}\rangle_0$. *If* H *is a derivation homotopy* $f \simeq g$ *then* $h = Ht + \mathbb{1}$ *is a twisting homotopy* $ft \simeq gt$.

Analogously for maps $C' \rightrightarrows C \in \langle\mathrm{DGC}\rangle_0$.

Proof. This is a straightforward computation using (2.9). □

N° 2 The (co)bar construction

Let $C \in \langle\mathrm{DGC}\rangle_0$ *and* $A \in \langle\mathrm{DGA}\rangle_0$. We introduce two constructions, which some authors call the "loop algebra" of C, resp. the "classifying coalgebra" of A.

Cobar construction (2.20)

We define $\Omega C = (\mathsf{T}(s_{-1}\widetilde{C}), \nabla) \in \langle\mathrm{DGA}\rangle_0$ as follows. Let $t : C \to \Omega C$ be the degree -1 map $t = i^1\langle\mathrm{id}\rangle q : C \to \widetilde{C} \to s_{-1}\widetilde{C} \subset \Omega C$. By (2.16) there are unique degree -1 derivations d, ∂ on ΩC with $dt = -td_C$ and $\partial t = t \cup t$ $(= m_{\Omega C}(t \otimes t)\Delta_C)$. One easily verifies $dd = 0 = \partial\partial$, $\partial d + d\partial = 0$ and $\varepsilon d = 0 = \varepsilon\partial$ (note first that it suffices to show $\partial^2 t = 0$ etc.). Hence the derivation $\nabla = d + \partial$ satisfies $\nabla\nabla = 0$, $\varepsilon\nabla = 0$. Thus $\Omega C \in \langle \mathrm{DGA}\rangle_0$. By definition $t = t_C : C \to \Omega C$ is a twisting map.

More explicitly, the *internal differential* d and the *external differential* ∂ on ΩC are the derivations determined by $d^1 = \langle -d_C\rangle : s_{-1}\widetilde{C} \to s_{-1}\widetilde{C}$ and $\partial^1 = \langle \mathrm{id}\rangle^{\otimes 2}\langle\widetilde{\Delta}\rangle : s_{-1}\widetilde{C} \to \widetilde{C}^{\otimes 2} \to (s_{-1}\widetilde{C})^{\otimes 2}$. Note that $\langle \mathrm{id}\rangle^{\otimes 2}$ throws in a sign $(-1)^k$ on $\widetilde{C}_k \otimes \widetilde{C}_\ell$, and that $(s_{-1}\widetilde{C})^{\otimes 2} \subset \mathsf{T}s_{-1}\widetilde{C}$ *via* m.

Bar construction (2.21)

We define $BA = (\mathsf{T}'(s\overline{A}), \nabla) \in \langle \mathrm{DGC}\rangle_0$ as follows. Let $t : BA \to A$ be the degree -1 map $t = j\langle \mathrm{id}\rangle p^1 : BA \to s\overline{A} \to \overline{A} \subset A$. Since BA is cocomplete (2.15) there are unique coderivations d, ∂ on BA with $td = -d_A t$, $t\partial = t \cup t$. One checks $dd = 0 = \partial\partial$, $d\partial + \partial d = 0$, $d\eta = 0 = \partial\eta$. Hence $\nabla = d + \partial$ has the desired properties. Further $t = t_A$ is a twisting map.

Theorem (2.22)

(a) *The mapping* $f \mapsto ft_C$ *is a bijection* $\mathrm{DGA}(\Omega C, A)_0 = \mathrm{T}(C, A)$. — *Given* $f, g \in \mathrm{DGA}(\Omega C, A)_0$ *the mapping* $H \mapsto Ht_C + \mathbb{1}$ *is a bijection* {derivation homotopies: $f \simeq g$} = {twisting homotopies: $ft_C \simeq gt_C$}.

(b) *Assume* C *cocomplete or* $A \in \langle \mathrm{dgA}\rangle_0$ *or* $A \in \langle \mathrm{dg^*A}\rangle_2$. *Then the mapping* $f \mapsto t_A f$ *is a bijection* $\mathrm{DGC}(C, BA)_0 = \mathrm{T}(C, A)$. — *Given* $f, g \in \mathrm{DGC}(C, BA)_0$ *the mapping* $H \mapsto t_A H + \mathbb{1}$ *is a bijection* {coderivation homotopies: $f \simeq g$} = {twisting homotopies: $t_A f \simeq t_A g$}.

Proof. Ad (a). Since $v \in \mathrm{T}(C, A)$ factors through $\widetilde{C} \to \overline{A}$ there is a unique map $f : \Omega C \to A \in \langle \mathrm{GA}\rangle_0$ with $ft_C = v$ (2.16). To show $D(f) = 0$ or equivalently (2.17), $D(f)t_C = 0$, is immediate by (2.9) (i). — Now let $h : ft_C \simeq gt_C$. Again by (2.16) there is a unique (f, g)-derivation $H : \Omega C \to A$ with $Ht_C = h - \mathbb{1}$. By (2.17) it remains to verify $D(H)t_C = ft_C - gt_C$. This is straightforward using (2.9) (ii). — The proof of (b) is analogous, the extra assumption guarantees that (2.16) (b) can be applied. □

In particular, $t_C : C \to \Omega C$ (resp. $t_A : BA \to A$) is a *universal twisting map* for C (resp. A). Moreover, the functors

$$\Omega : \langle \mathrm{DGC}\rangle_0 \rightleftarrows \langle \mathrm{DGA}\rangle_0 : B \tag{2.23}$$

are homotopy preserving; and they are adjoint when we restrict to cocomplete objects in $\langle \mathrm{DGC}\rangle_0$.

Twisted tensor products (2.24)

Given $v \in \mathrm{T}(C, A)$ let $\partial_v : C \otimes A \to C \otimes A$ be the map $(1_C \otimes m_A)(1_C \otimes v \otimes 1_A)(\Delta_C \otimes 1_A)$. One checks that $C \otimes A$ with $d_\otimes - \partial_v$ is a chain complex, where $d_\otimes$ denotes the tensor product differential of $C \otimes A$ ([GMu], [May]). It is denoted by $C \otimes_v A = (C \otimes A, d_\otimes - \partial_v)$. This construction is natural in C and A.

Let $t : C \to \Omega C$ (resp. $BA \to A$) be the universal twisting map. One can show that $C \otimes_t \Omega C$ resp. $BA \otimes_t A$ is acyclic rel K (§ 3, N° 2). The classical formulae (see e.g. [AdHi] resp. [LaS]) can be liberated from elements.

N° 3 Compatibility with tensor product

Let $C, C' \in \langle \mathrm{DGC} \rangle_0$ *and* $A, A' \in \langle \mathrm{DGA} \rangle_0$. Given maps $v : C \to A$, $w : C' \to A'$ of the same degree, we define there *folding* $v * w : C \otimes C' \to A \otimes A'$ by $v * w = 1\!\!1 \otimes w + v \otimes 1\!\!1$. If v, w are twisting maps, so is $v * w$.

Using this we define a natural map

$$j : \Omega(C \otimes C') \longrightarrow \Omega C \otimes \Omega C' \qquad \text{in } \langle \mathrm{DGA} \rangle_0 \tag{2.25}$$

by $j\, t_{C \otimes C'} = t_C * t_{C'}$. In a dual manner we define

$$\tilde{\jmath} : BA \otimes BA' \longrightarrow B(A \otimes A') \qquad \text{in } \langle \mathrm{DGC} \rangle_0 \tag{2.26}$$

by $t_{A \otimes A'} \tilde{\jmath} = t_A * t_{A'}$. Here we use (2.15) that $BA \otimes BA'$ is cocomplete. It is easily verified that j and $\tilde{\jmath}$ are commutative monoidal transformations (B.2).

Shuffle (co)multiplication (2.27)

Let A be *commutative*, i.e. the multiplication m of A satisfies $\tau m = m$, cf. (B.1). Then $m : A \otimes A \to A$ is an augmented DGA-map. Since $\tilde{\jmath}$ is a commutative monoidal transformation,

$$\nu \; : \; BA \otimes BA \xrightarrow{\tilde{\jmath}} B(A \otimes A) \xrightarrow{B(m)} BA \qquad \text{in } \langle \mathrm{DGC} \rangle_0$$

is a unital, associative, commutative multiplication; i.e. (BA, ν) is a *DG Hopf coalgebra*. Observe that ν is unique with the property $t\nu = m(t * t)$ where $t = t_A$. One checks that $m(t * t) = t(\pi_1 + \pi_2)$ where $\pi_1 = \rho(1_{BA} \otimes \varepsilon)$ and $\pi_2 = \lambda(\varepsilon \otimes 1_{BA}) : BA \otimes BA \to BA$. Hence different multiplications on A yield the same ν. In fact, ν is defined on any tensor coalgebra, since $\mathsf{T}'V = B(K \oplus s_{-1}V)$ where $m = 0$ on $s_{-1}V$. Hence the above says that $\nu : BA \otimes BA \to BA$ is a chain map if A is commutative. There is an explicit formula for ν ([EMc]) which is not needed here; since shuffle permutations appear in it, ν is called the *shuffle multiplication* of BA.

Dually there is a *shuffle diagonal* $\varphi : \Omega C \to \Omega C \otimes \Omega C$ in case C is cocommutative. This diagonal is counital, coassociative and cocommutative, i.e. $(\Omega C, \varphi)$ is a DG Hopf algebra.

N° 4 Homological properties

We briefly discuss the classical homological properties of the (co)bar construction. Whenever we speak of homology (e.g. H_*-equivalence) we assume (in addition) that $\mathcal{V}$ is abelian.

First let us consider the adjoint functors $\Omega : \langle \mathrm{dgC} \rangle_1 \rightleftarrows \langle \mathrm{dgA} \rangle_0 : B$. These restrictions of (2.23) enjoy the following properties.

THEOREM (2.28)

(a) *The adjunction maps* $\Omega B(A) \to A$ *and* $C \to B\Omega(C)$ *are chain homotopy equivalences (thus* H_**-equivalences).*

(b) *The maps j and $\tilde{\jmath}$ are chain homotopy equivalences (thus H_*-equivalences).*

(c) *B preserves H_*-equivalences, and so does Ω when restricted to $\langle \mathrm{dgC} \rangle_2$.*

Proof. In the case of modules the standard reference for (a)–(b) is [HMS | p. 148, 176]. A sufficiently general ad hoc argument for (a) is given in [Pr], and in [LaS | p. 367] there are simple formulae showing that j and $\tilde{\jmath}$ are part of an EZ-morphism (§ 4). Part (c) follows from [EM] applied to the spectral sequence associated with the standard filtration (2.12) of BA resp. ΩC. In fact, for BA this filtration is always bicomplete (2.15), while for ΩC this is the case if $C \in \langle \mathrm{dgC} \rangle_2$. □

The adjoint functors

$$(*) \qquad \Omega : \langle \mathrm{dgC} \rangle_2 \rightleftarrows \langle \mathrm{dgA} \rangle_1 : B$$

are of main interest here. If these are considered, one can prove (a)–(c) also by using the comparison theorem for spectral sequences, applied to the spectral sequence of a twisted tensor product; cf. proof of (4.10) (c).

We do not know whether the extra assumption for Ω in (c) is really needed.

Dualization (2.29)

The adjoint functors $\Omega : \langle \mathrm{dg^*C} \rangle_1 \rightleftarrows \langle \mathrm{dg^*A} \rangle_2 : B$ both preserve H^*-equivalences and satisfy (a)–(b) by similar arguments. In fact, we have an equivalence to the mentioned properties of $(*)$ by categorical duality.

It seems the same holds for the functors $\Omega : \langle \mathrm{dg^*C} \rangle_0 \rightleftarrows \langle \mathrm{dg^*A} \rangle_1 : B$. For adjointness, thus for (a), assume here that coalgebras are cocomplete. This problem is categorically dual to the variant of (2.28) where one restricts to $\langle \text{complete dgA} \rangle_0$ and replaces Ω by $\widehat{\Omega}$ (the dual of B). Anyway, we do not need this here.

§ 3 Acyclic Models

N° 1 Representable functors

We assume that $\mathcal{V}$ (our fixed additive category) has arbitrary direct sums. We do not assume here that $\mathcal{V}$ is an additive monoidal category, neither that $\mathcal{V}$ has (co)kernels.

Let $\mathcal{T}$ be a category and $\mathcal{M} \subset \mathcal{T}$ a set of objects (the "models"). Let $F : \mathcal{T} \to \mathcal{V}$ be a functor. Define a new functor $\check{F} : \mathcal{T} \to \mathcal{V}$ by

$$(2.30) \qquad \check{F}(X) = \bigoplus_{M \in \mathcal{M}} \mathcal{T}(M, X) \otimes F(M)$$

where by definition $S \otimes V = \bigoplus_{\sigma \in S} V$ for any set S and $V \in \mathcal{V}$. Then for any $M \in \mathcal{M}$ and $\sigma \in \mathcal{T}(M, X)$ we have the inclusion $i_\sigma : F(M) \to \check{F}(X)$. For $f : X \to Y \in \mathcal{T}$ define $\check{F}(f)$ by $\check{F}(f) i_\sigma = i_{f\sigma}$. One checks that $\check{F}$ is indeed a functor. Define a natural map

$$\Phi : \check{F} \to F , \qquad \Phi_X i_\sigma = F(\sigma) : F(M) \to F(X) .$$

Let $F, G : \mathcal{T} \rightrightarrows \mathcal{V}$ be functors and $f : F \to G$ a natural map, define

$$\check{f} : \check{F} \to \check{G}\,, \qquad \check{f}_X i_\sigma = i_\sigma f_M\,.$$

One checks that $\Phi^{(G)}\check{f} = f\Phi^{(F)}$. Further $(fg)\check{} = \check{f}\check{g}$ and $\check{1} = 1$. Finally $f \mapsto \check{f}$ is an abelian group map.

For a reformulation see proof of (2.39).

Definition (2.31)

Let $\mathcal{T}$ be a category with models $\mathcal{M}$. A functor $F : \mathcal{T} \to \mathcal{V}$ is *representable* with models $\mathcal{M}$ if there is a natural map $\Psi : F \to \check{F}$ with $\Phi^{(F)}\Psi = \mathrm{id}_F$.

Remark. Some authors (e.g. [May]) use a more general notion of representable functor in the case of modules over some ring, i.e. in (2.30) they replace $F(M)$ by the free module generated by $F(M)$. However this generality is unconvenient in theory and seems useless in practice.

LEMMA (2.32)

Let $F, G : \mathcal{T} \rightrightarrows \mathcal{V}$ and suppose F is representable. Then given $f, g : F \rightrightarrows G$ we have: $f = g \Leftrightarrow f_M = g_M$ for $M \in \mathcal{M}$. □

In fact, the relation $f = f\Phi\Psi = \Phi\check{f}\Psi$ shows that a natural map $F \to G$ is uniquely determined by what it does on models.

LEMMA (2.33)

Let $F : \mathcal{T} \to \mathcal{V}$ be representable with models $\mathcal{M}$.

(a) *If $H : \mathcal{V} \to \mathcal{W}$ preserves direct sums, then $H \circ F$ is also representable with models $\mathcal{M}$.*

(b) *If $\mathcal{M} \subset \mathcal{M}'$ then F is also representable with models $\mathcal{M}'$.*

(c) *Let $E : \mathcal{T} \to \mathcal{V}$ be any functor and $\alpha : E \rightleftarrows F : \beta$ natural maps with $\beta\alpha = 1_E$. Then also E is representable with models $\mathcal{M}$.* [Put $\Psi^{(E)} = \check{\beta}\Psi^{(F)}\alpha : E \to \check{E}$.] □

THEOREM (2.34)

Let $A : \mathcal{T} \to \mathcal{V}$, $B : \mathcal{T}' \to \mathcal{V}'$ be representable functors with models $\mathcal{M}$ resp. $\mathcal{M}'$. Suppose $\Gamma : \mathcal{V} \times \mathcal{V}' \to \mathcal{W}$ preserves direct sums in each variable. Then $\Gamma \circ (A \times B) : \mathcal{T} \times \mathcal{T}' \to \mathcal{W}$ is representable with models $\mathcal{M} \times \mathcal{M}'$. □

THEOREM (2.35)

When $A, B : \mathcal{T} \rightrightarrows \mathcal{V}$ are representable with models $\mathcal{M}$, so is $(A \times B) \circ \Delta : \mathcal{T} \to \mathcal{V} \times \mathcal{V}$. □

Here $\Delta : \mathcal{T} \to \mathcal{T} \times \mathcal{T}$ is the diagonal functor. By (2.33) (a) applied to $H = \oplus$ we obtain: *$A \oplus B : \mathcal{T} \to \mathcal{V}$ is representable with models $\mathcal{M}$.*

THEOREM (2.36)

Let $W : \mathcal{T} \rightleftarrows \mathcal{U} : \nabla$ be adjoint functors (W left adjoint). Let $F : \mathcal{T} \to \mathcal{V}$ be representable with models $\mathcal{M}$. Then $F\nabla$ is representable with models $W(\mathcal{M})$. □

The proofs are straightforward.

We need the following special case of the last theorem. Let $P \in \mathcal{T}$ be any object. Consider the category $\mathcal{T}^P$ of maps $\eta : P \to X \in \mathcal{T}$ written (X, η). Denote by $\phi : \mathcal{T}^P \to \mathcal{T}$ the obvious functor $(X, \eta) \mapsto X$. Assuming that $\mathcal{T}$ has finite coproducts, ϕ has the left adjoint $Y \mapsto Y^+ = (Y \sqcup P, \text{in}_2)$. Hence (2.36) implies ...

Corollary (2.37)

Let $F : \mathcal{T} \to \mathcal{V}$ be representable with models $\mathcal{M}$. Then $F\phi : \mathcal{T}^P \to \mathcal{V}$ is representable with models $\{M^+\}_{M \in \mathcal{M}}$ where $M^+ = (M \sqcup P, \text{in}_2)$. □

Examples (2.38)

(a) Let (only here) $C : \mathcal{S} \to \langle \text{dg} \rangle$ denote the *non-normalized* chain functor, cf. App. B, N° 2. Then C_n is representable with the single model $\Delta[n]$ hence with models $\mathcal{M} = \{\Delta[p]\}_{p \geqslant 0}$. Define $\Psi_X : C_n(X) \to X^n \otimes C_n(\Delta[n])$ by (e.g.) $\sigma \mapsto \sigma \otimes e_n$ where $e_n = (0, \dots, n) \in \Delta[n]$.

(b) Regard the restriction $C : \mathcal{S}_1 \to \langle \text{dg} \rangle$. Here C_n is representable with model $\overline{\Delta}[n]$, where $\overline{X} = X/X^0$ for spaces X. This follows from (2.36) since $X \mapsto \overline{X}$ is left adjoint to $\mathcal{S}_1 \hookrightarrow \mathcal{S}$.

(c) The degree n component of $C(X) \otimes C(Y)$ as a functor on $\mathcal{S} \times \mathcal{S}$ is representable with models $\{(\Delta[p], \Delta[q])\}_{p+q \leqslant n}$, by (a) and (2.34), (2.35). The restriction to $\mathcal{S}_1 \times \mathcal{S}_1$ is thus representable with models $\{(\overline{\Delta}[p], \overline{\Delta}[q])\}_{p+q \leqslant n}$.

(d) These statements remain true when C denotes (as usual) the *normalized* chain functor. This is clear by (2.33) (c).

N° 2 The method of acyclic models

Let $C \in \langle \text{DG} \rangle$ and suppose there are given: a chain complex V, chain maps $\iota : V \rightleftarrows C : \rho$ with $\rho\iota = 1$ and a degree $+1$ map $\xi : V \to V$, such that $d\xi + \xi d = 1 - \iota\rho$. Then C is said to be *acyclic* rel $\rho : C \to V$. We are interested in the case where V is concentrated in a single degree.

The following lemma provides the "method of acyclic models".

Lemma (2.39)

Let $F : \mathcal{T} \to \mathcal{V}$ be representable with models $\mathcal{M}$, and let $G : \mathcal{T} \to \langle \text{DG} \rangle$ be such that $G(M)$ is acyclic rel $\rho^M : G(M) \to V^M$ $(M \in \mathcal{M})$. Let $g : F \to G_n$ be a natural map with $d_M g_M = 0$, $\rho^M g_M = 0$ $(M \in \mathcal{M})$. Then there is a natural map $g' : F \to G_{n+1}$ with $dg' = g$.

We first give a reformulation of the problem entirely in terms of the additive category $\mathcal{V}^{\mathcal{T}}$ of functors $\mathcal{T} \to \mathcal{V}$ and natural maps (it is however not necessary to use this category explicitly). The rest of the proof is a simple exercise. The mentioned reformulation can also be used in various applications of the method of acyclic models, i.e. regard *functors* into an additive category as *objects* of another additive category.

Proof. Recall that (given $\mathcal{M}$) we have the additive functor $\check{}\ : \mathcal{V}^{\mathcal{T}} \to \mathcal{V}^{\mathcal{T}}$ and a natural transformation $\Phi : \check{}\ \to \text{Id}$. Recall that the representability of $F \in \mathcal{V}^{\mathcal{T}}$ means there is $\Psi : F \to \check{F} \in \mathcal{V}^{\mathcal{T}}$ with $\Phi^{(F)}\Psi = \text{id}_F$.

We show that actually a few data are sufficient to obtain objects resp. morphisms of $\mathcal{V}^{\mathcal{T}}$. As in (2.30) any family $\{A^M\}_{M \in \mathcal{M}}$ of objects of $\mathcal{V}$ yields $\bar{A} \in \mathcal{V}^{\mathcal{T}}$, and any family of maps $\alpha = \{\alpha^M : A^M \to B^M\}$ in $\mathcal{V}$ yields a natural map $\bar{\alpha} : \bar{A} \to \bar{B} \in \mathcal{V}^{\mathcal{T}}$. This gives a functor $\bar{}\ : \mathcal{V}^{[\mathcal{M}]} \to \mathcal{V}^{\mathcal{T}}$ where $[\mathcal{M}]$ is the discrete category with objects $\mathcal{M}$. This functor is additive and front composition with the obvious additive functor $\mathcal{V}^{\mathcal{T}} \to \mathcal{V}^{[\mathcal{M}]}$ (given by restriction) yields the functor

$\check{}: \mathcal{V}^{\mathcal{T}} \to \mathcal{V}^{\mathcal{T}}$ considered above, i.e. $\bar{F} = \check{F}$, $\bar{g} = \check{g}$ etc. We write $r_n = \bar{\rho}_n : \check{G}_n \to \bar{V}_n$, $i_n = \bar{\iota}_n : \bar{V}_n \to \check{G}_n$, $x_n = \bar{\xi}_n : \check{G}_n \to \check{g}_{n+1}$.

By assumption $\check{G}, \check{d}$ (regarded as a chain complex over $\mathcal{V}^{\mathcal{T}}$) is acyclic rel $\bar{V}$ via r, i, x. Further $\check{g} : \check{F} \to \check{G}_n$ satisfies $\check{d}\check{g} = 0$, $r\check{g} = 0$. Put $g' = \Phi x \check{g} \Psi$, then since F is representable $dg' = d\Phi x\check{g}\Psi = \Phi \check{d} x \check{g}\Psi = \Phi(1 - ir - x\check{d})\check{g}\Psi = \Phi\check{g}\Psi = g\Phi\Psi = g$ as desired. □

Definitions (2.40)

An *augmentation* for a functor $G : \mathcal{T} \to \langle \mathrm{dg} \rangle$ is a natural chain map $\varepsilon : G \to K$ where $K \in \langle \mathrm{dg} \rangle$ is concentrated in degree 0 (regard K as a constant functor). We say G is *acyclic on models* rel $\varepsilon : G \to K$ if $G(M)$ is acyclic rel $\varepsilon : G(M) \to K$ for all $M \in \mathcal{M}$.

THEOREM (2.41)

Let $F, G : \mathcal{T} \rightrightarrows \langle \mathrm{dg} \rangle$ be functors with augmentation. Suppose that F_n is representable for all n, and that G is acyclic on models rel $\varepsilon : G \to K$.

(a) *There is a natural chain map $f : F \to G$ with $\varepsilon f = \varepsilon$.*

(b) *Given natural chain maps $f, g : F \rightrightarrows G$ with $\varepsilon f = \varepsilon = \varepsilon g$ there is a natural chain homotopy $h : F \to G$ from f to g.*

ADDENDUM (2.42)

Suppose, in addition, that F_n is representable with models $\mathcal{M}^n \subset \mathcal{M}$, and that $G_{n+1}(M) = 0$ for all $M \in \mathcal{M}^n$. Then $f : F \to G$ with $\varepsilon f = \varepsilon$ exists uniquely.

Proof. The theorem is an easy consequence of (2.39). For the addendum, note that by assumption $\varepsilon : G_0(M) \cong K$ for $M \in \mathcal{M}^0$, and $\xi^M : G_{n-1}(M) \rightleftarrows G_n(M) : d$ satisfy $\xi^M d = 1$ for $M \in \mathcal{M}^n$ $(n \geqslant 1)$. The uniqueness of f follows by induction, since f_n is determined by what it does on $\mathcal{M}^n$, see (2.32). □

Examples (2.43)

(a) The normalized chain functor $C : \mathcal{S} \to \langle \mathrm{dg} \rangle$ is acyclic on models $\{\Delta[p]\}_{p \geqslant 0}$ rel $\varepsilon : C \to K$ (the canonical augmentation). This follows from the corresponding well-known property of the chain functor $X \mapsto KX$. For the definition of these functors in the general setting, see App. B, N° 2. A similar property is derived for $C(X) \otimes C(Y)$ on $\mathcal{S} \times \mathcal{S}$.

(b) At least in case $\mathcal{V}$ is abelian one checks the following criterion. Let $C \in \langle \mathrm{dg} \rangle$ be projective in each degree, and suppose $H_0(C)$ is also projective while $H_{>0}(C) = 0$. Then C is acyclic rel $H_0(C)$.

N° 3 Duality

There is a duality between cochain complexes over $\mathcal{V}$ and chain complexes over $\mathcal{V}^{\mathrm{op}}$. This duality is *not* in general compatible with the graded tensor product, because in its definition infinite direct sums are involved. Hence in this sense the dual of a DGA is not in general a DG*C. However this holds if the DGA is bounded below. We study more generally adjoint additive cofunctors, which also applies to functional duality (Hom-cofunctors), see examples (2.57). Before doing so we introduce the dual of the notion of representable functor, i.e. 'representable' cofunctor. In [Bou G] the authors use the term 'corepresentable', but as we always explicitly distinguish between functors and cofunctors, there is no ambiguity here. A justification of this terminology is given in App. B, N° 3.

Definition (2.44)

A cofunctor $F : \mathcal{T} \to \mathcal{V}$ is *representable* if the obvious natural map $\Phi : F \to \hat{F}$ has a natural retraction Ψ where

$$\hat{F}(X) = \prod_{M \in \mathcal{M}} F(M)^{\mathcal{T}(M,X)}$$

with $V^S = \operatorname{Hom}(S, V) := \prod_{s \in S} V$ for any set S and $V \in \mathcal{V}$. We have to assume here that products exist in $\mathcal{V}$.

Let $\mathcal{V}$, $\mathcal{X}$ be additive categories. Suppose we are given *adjoint additive cofunctors*

(2.45) $$\# : \mathcal{V} \rightleftarrows \mathcal{X} : {}^{\vee}$$

i.e. there is a natural group isomorphism $\mathcal{V}(V, X^{\vee}) \approx \mathcal{X}(X, \#V)$ which we denote by $f \mapsto f'$ in both directions.

This pair of cofunctors resp. isomorphism is fixed throughout this subsection. Since adjoint cofunctors carry colimits to limits, we have the following ...

Lemma (2.46)

Let $\mathcal{T}$ be a category with models $\mathcal{M}$. If $F : \mathcal{T} \to \mathcal{X}$ is a representable functor, then $F^{\vee} : \mathcal{T} \to \mathcal{V}$ is a representable cofunctor. □

Here we write $F^{\vee} = {}^{\vee} \circ F$. The converse statement holds provided the given cofunctor ${}^{\vee}$ is a co-equivalence of categories.

We define "extended" additive cofunctors

(2.47) $$\# : \langle \mathrm{DG}^* \rangle_{\mathcal{V}} \rightleftarrows \langle \mathrm{DG} \rangle_{\mathcal{X}} : {}^{\vee}$$

as follows: As a graded object $(\#E)_n = \#(E^n)$. For any graded map $g : F \to E$ (of degree k) let $\#g$ (of degree $-k$) be given by $(\#g)_n = (-1)^{kn} \#(g^{n-k}) : \#(E^n) \to \#(F^{n-k})$. Then we define the differential of $\#E$ to be $-\#(d_E)$. Analogously the adjoint cofunctor is extended to chain complexes and graded maps.

The extended cofunctors are adjoint. In fact, we have the natural graded isomorphism

(2.48) $$\operatorname{Hom}(C, \#E) = \operatorname{Hom}(E, C^{\vee})$$

written as $f \mapsto f'$ (in both directions) and defined as follows: Given $f : C \to \#E$ of degree k let $(f')^{n+k} = (-1)^{kn} (f_n)' : E^{n+k} \to (C_n)^{\vee}$. Recall that $\operatorname{Hom}_k(E, C^{\vee})$ are the maps of (upper) degree $-k$. One checks the formulas

(2.49) $$(\#g \circ f)' = (-1)^{|f|\,|g|} f' \circ g\,, \quad (f \circ h)' = (-1)^{|f|\,|h|} h^{\vee} \circ f'$$

where $f : C \to \#E$, $g : F \to E$, $h : B \to C$ are graded maps. Hence (2.48) is an isomorphism of chain complexes:

$$D(f') = (-d_C^{\vee}) f' - (-1)^{|f|} f' d_E = (-(-1)^{|f|} f d_C + (-\#d_E) f)' = D(f)'\,.$$

It follows that if $g : C \to \#E$ is a chain map then g' is a cochain map, and similarly for homotopies.

LEMMA (2.50)

If $C \in \langle \mathrm{DG} \rangle_{\mathcal{X}}$ is acyclic rel $\rho : C \to V$ then $C^{\vee}$ is acyclic rel $\rho^{\vee} : V^{\vee} \to C^{\vee}$. □

Here the notion of cochain complex 'acyclic rel subcomplex' is essentially the same as for chain complexes. The converse statement holds provided the given cofunctor $^{\vee}$ is a co-equivalence of categories, since then so is the extended $^{\vee}$.

Let $\mathcal{V}$ resp. $\mathcal{X}$ be additive monoidal with unit objects K resp. J. Moreover, suppose there are given maps in $\mathcal{X}$

$$\phi : \#E \otimes \#F \to \#(E \otimes F)\,, \qquad u : J \to \#K \tag{2.51}$$

the former being natural under $\mathcal{V} \times \mathcal{V}$. By adjunction we obtain $v = u' : K \to J^{\vee}$ and a unique map $\psi : X^{\vee} \otimes Y^{\vee} \to (X \otimes Y)^{\vee}$ such that the following diagram commutes for all $f : E \to C^{\vee}$, $g : F \to B^{\vee}$:

$$\begin{array}{ccc} & C \otimes B & \\ {}^{f' \otimes g'}\swarrow & & \searrow^{(\psi(f \otimes g))'} \\ \#E \otimes \#F & \xrightarrow{\ \phi\ } & \#(E \otimes F)\,. \end{array} \tag{2.52}$$

Consider the adjunction (2.47) and the graded tensor product in these categories. We now define chain maps ϕ, u extending the given maps (2.51). For u there is nothing to do. Given $E,\, F \in \langle \mathrm{DG}^{*} \rangle_{\mathcal{V}}$ define $\phi : \#E \otimes \#F \to \#(E \otimes F)$ as follows: the component $\phi^{i,j} : (\#E)_i \otimes (\#F)_j \to (\#(E \otimes F)\,)_{i+j}$ is the composite of $(-1)^{ij}\phi : \#(E^i) \otimes \#(F^j) \to \#(E^i \otimes F^j)$ followed by the $\#$-image of the natural projection $(E \otimes F)^{i+j} \to E^i \otimes F^j$. In the same way we extend ψ to a natural chain map. One checks:

LEMMA (2.53)

Diagram (2.52) commutes for any graded maps f, g.

If the given ϕ is a natural isomorphism, so is the extended ϕ provided E, F are bounded below.

If the given ϕ is associative, commutative, resp. unital, so is the extended ϕ. □

Definition - Remarks (2.54)

Here ϕ is *associative* if $\phi(\phi \otimes 1) = (\#\alpha)\phi(1 \otimes \phi)\alpha$, *commutative* if $(\#\tau)\phi = \phi\tau$, and *unital* (with u) if $\phi(u \otimes 1) = (\#\lambda)\lambda$ and analogously with ρ, cf. (B.2). When $\#$ has an adjoint as above, any of these properties is equivalent to the corresponding property of (ψ, v).

Recall that the (graded) associativity isomorphism α is defined for bounded below graded objects, resp. for all graded objects if the given tensor products are *strongly* additive. The associativity of the given ϕ implies the same for the extended ϕ whenever this makes sense. If it holds we obtain a well-defined chain map $\phi : (\#E)^{\otimes n} \to \#(E^{\otimes n})$, and the obvious generalization of (2.52) for tensor products with more than two factors.

Let ϕ be associative and unital. Let u and v be isomorphisms.

One checks that then $\#$ (and analogously $^{\vee}$) yields a cofunctor $\# : \langle \mathrm{DG}^{*}\mathrm{C} \rangle_0 \to \langle \mathrm{DGA} \rangle_0$ where $\#E$ receives the structure $m = \#\Delta \circ \phi$, $\eta = \#\varepsilon u$, $\varepsilon = u^{-1}\#\eta$. Given $E, F \in \langle \mathrm{DG}^{*}\mathrm{C} \rangle_0$ one checks that ϕ is a map in $\langle \mathrm{DGA} \rangle_0$.

Now let $C \in \langle\mathrm{DGC}\rangle_0$ and $E \in \langle\mathrm{DG^*C}\rangle_0$ so that $C^\vee \in \langle\mathrm{DG^*A}\rangle_0$ and $\#E \in \langle\mathrm{DGA}\rangle_0$. Using (2.52) one easily sees that (2.48) is compatible with the cup product: $(f \cup g)' = f' \cup g'$ and $\mathbb{1}' = \mathbb{1}$. Hence if $v : C \to \#E$ is a twisting map, so is $v' : E \to C^\vee$. In fact, $D(v') = D(v)' = (v \cup v)' = v' \cup v'$ and $v'\eta = (\#\eta \circ v)' = 0$, $\varepsilon v' = \eta^\vee v' = (v\eta)' = 0$.

Together with a similar observation on twisting homotopies we obtain ...

Theorem (2.55)
The mapping $v \mapsto v'$ *is a bijection* $\mathrm{T}(C, \#E) = \mathrm{T}(E, C^\vee)$. *Moreover,* $v \simeq w \Leftrightarrow v' \simeq w'$. *In fact, the mapping* $h \mapsto h'$ *is a bijection* $\{$twisting homotopies: $v \simeq w\} = \{$twisting homotopies: $v' \simeq w'\}$. □

Definition (2.56)
The cofunctor $\# : \mathcal{V} \to \mathcal{X}$ together with ϕ, u is a *co-equivalence of monoidal categories*, if the following holds:

(i) $\#$ is a co-equivalences of categories,

(ii) ϕ is a natural isomorphism,

(iii) u is an isomorphism,

(iv) ϕ is associative, commutative and unital.

Note that in this situation we have a perfect duality between algebras over $\mathcal{V}$ and coalgebras over $\mathcal{X}$, and vice versa.

Examples (2.57)
(a) Let $\mathcal{X} = \mathcal{V}^{\mathrm{op}}$ and (2.45) the obvious co-isomorphism. The tensor product in $\mathcal{V}^{\mathrm{op}}$ is defined as in $\mathcal{V}$, i.e. $\# : \mathcal{V} \to \mathcal{V}^{\mathrm{op}}$ with $\phi = \mathrm{id}$, $u = \mathrm{id}$ is a co-isomorphism of monoidal categories, in particular (i)–(iv) are satisfied. We obtain a co-isomorphism between the respective monoidal categories of *bounded below* complexes. Such a restriction is necessary, since $\mathcal{V}^{\mathrm{op}}$ is not strongly additive (even if $\mathcal{V}$ is so). This example is used for the purpose of *categorical dualization*. For a non-trivial application, see (2.65).

(b) Let $\mathcal{V} = \mathcal{X} = \mathcal{M}od_{\mathsf{K}}$ where K is a commutative ring. The *functional dualization* cofunctor $\mathrm{Hom}(-, \mathsf{K})$ is adjoint to itself, $\phi = \psi$ is the familiar map $\phi(\alpha \otimes \beta)(x \otimes y) = \alpha(x)\beta(y)$ and $u = v$ the obvious isomorphism.— Here the extended cofunctors coincide up to the identification of chain complexes with cochain complexes, and $\#E = \mathrm{Hom}(E, \mathsf{K})$ is the chain complex of graded maps (2.2). The extended ϕ is given by the same formula as the given ϕ except for a factor $(-1)^{|\beta||x|}$. The graded adjunction $f \mapsto f'$ is given by $f'(x)(y) = (-1)^{|x|\,|y|} f(y)(x)$. Thus the various signs used in the definitions above respect standard conventions; it seems one has to use these signs to obtain (2.55).— This example is of importance in Chp. 4, § 2. Notice that (iii)–(iv) are satisfied, but (i)–(ii) are not (unless we restrict to the full subcategory of finitely generated free modules). In particular, we have (2.55) in full generality.

(c) There is an analogous story with adjoint *functors* (instead of cofunctors). In this situation the signs in the various formulas must be omitted. Using (a) we obtain immediately analogous results.

N° 4 Acyclic model theorems for twisting maps

Let $\mathcal{V}$ be a strongly additive monoidal category. Using the method of acyclic models we establish results on the existence and homotopy uniqueness of natural twisting maps. The first result deals with representable *co*functors etc. (see N° 3). We assume thus that $\mathcal{V}$ has arbitrary products. Note that differentials and twisting maps have 'upper' degree +1.

Bigraded coalgebras (2.58)

Let $B_*^* = \{B_n^m\}_{n,m \geqslant 0}$ be a *bigraded coalgebra* over $\mathcal{V}$. Since bigraded objects (= graded graded objects) form an additive monoidal category in an obvious way this notion is defined. The diagonal Δ is thus given by maps (of upper degree 0)

$$\Delta_n = \sum \Delta_{n,n-k} : B_n \to \bigoplus_{k=0}^n B_k \otimes B_{n-k}$$

where $B_n = B_n^*$ and as usual it is assumed that this diagonal is coassociative and has a counit $\varepsilon : B_*^* \to K$ (a map of bi-degree 0). In addition we assume that $B_0 = K$ and that the counit is the identity on B_0^0. Finally we assume there are given bigraded maps d, δ (of upper degree +1 and lower degree 0 resp. −1) which are coderivations satisfying $dd = 0$, $\delta\delta = 0$ and $d\delta + \delta d = 0$. Hence we have for all $n \geqslant 0$ maps (of upper degree +1)

$$d_n : B_n \to B_n \qquad \delta_n : B_n \to B_{n-1} .$$

It is clear that if B_*^* is as above, then $B = \bigoplus_{n=0}^\infty B_n$ is an object of $\langle \mathrm{DG^*C} \rangle_0$ with differential $\nabla = d + \delta$ and obvious coaugmentation. In fact d, δ are well-defined differentials (and coderivations) on B satisfying $d\delta + \delta d = 0$, hence ∇ is a differential (and coderivation).

Definitions (2.59)

An object $B \in \langle \mathrm{DG^*C} \rangle_0$ is Σ-*decomposable* if there is a decomposition $B = \bigoplus_{n=0}^\infty B_n$ such that B_*^* is a bigraded coalgebra as above inducing the structure of B. We call d the *internal* and δ the *external* differential of B_*^* resp. B.

In particular, we have a notion of Σ-decomposable cofunctor $B : \mathcal{T} \to \langle \mathrm{DG^*C} \rangle_0$. For example, the bar construction B is Σ-decomposable.

Moreover, we have the following result. Recall that if $B, C \in \langle \mathrm{DG^*C} \rangle_0$, so is $B \otimes C$ with diagonal $\Delta = \tau_{2,3}(\Delta_B \otimes \Delta_C)$, cf. (B.1).

Proposition (2.60)

If B, C are Σ-decomposable, so is $B \otimes C$ with $(B \otimes C)_n = \bigoplus_{k=0}^n B_k \otimes C_{n-k}$. □

Theorem (2.61)

Let the cofunctor $A : \mathcal{T} \to \langle \mathrm{dg^*A} \rangle_0$ be degreewise representable with models $\mathcal{M}$. Let the cofunctor $B : \mathcal{T} \to \langle \mathrm{DG^*C} \rangle_0$ be Σ-decomposable, such that $B_n(M), d$ is acyclic rel $\iota_n^M : V_n^M \to B_n(M)$ where $V_n^M \in \langle \mathrm{DG^*} \rangle$ is concentrated in degree $-n$ $(M \in \mathcal{M},\ n \geqslant 2)$.

(a) *Given $v_1 : B_1 \to A$ of degree +1, with $v_1 d_1 + d v_1 = 0$, $\varepsilon v_1 = 0$, and $v_1 \delta_2 = v_1 \cup v_1$ in degree −2, there is a twisting map $v : B \to A$ extending v_1.*

(b) *Given twisting maps $v, w : B \rightrightarrows A$ with $v_1 = w_1$, there is a twisting homotopy $h : v \simeq w$.*

Example (2.62)

The assumption in (a) that $v_1\delta = v_1 \cup v_1$ in degree -2 means that the following diagram commutes on B^2_{-2}.

$$(*) \qquad \begin{array}{ccccc} B_2 & \xrightarrow{\Delta_{1,1}} & B_1 \otimes B_1 & \xrightarrow{v_1 \otimes v_1} & A \otimes A \\ \downarrow{\scriptstyle \delta_2} & & & & \downarrow{\scriptstyle m} \\ B_1 & & \xrightarrow{\quad v_1 \quad} & & A \end{array}$$

In case $B = B(A')$ is a bar construction, $\Delta_{1,1}$ is an isomorphism and $\delta_2\Delta^{-1}_{1,1}$: $B_1 \otimes B_1 \to B_1$ is given by the multiplication of A'. The mentioned assumption is then equivalent to $f : A' \to A$ being multiplicative in degree 0, where f is the augmented chain map defined by $ft_{A'} = v_1$ and $f\eta = \eta$. In fact (a) states that such f extends to $B(A')$, in other words: f is *strongly homotopy multiplicative.* The first task in the proof of (a) is to construct a (degree $+1$) chain homotopy $v_2 : B_2 \to A$ for $(*)$, i.e. $D(v_2) = -v_1\delta_2 + v_1 \cup v_1$.

Remarks (2.63)

(a) Notice that the decomposition comes with maps $i_n : B_n \to B$ and induces obvious maps $p_n : B \to B_n$. These are chain maps with respect to the internal differential. Given a graded map $f : B \to A$, the restriction $f_k : B_k \to A$ is considered as a graded map $B \to A$ by putting it 0 on B_n for $n \neq k$. Let $f, g : B \rightrightarrows A$ be graded maps where $B \in \langle \mathrm{DG^*C}\rangle_0$ is Σ-decomposable and $A \in \langle \mathrm{DG^*A}\rangle_0$. Then

$$(f \cup g)_n = \textstyle\sum_{k=0}^{n} f_k \cup g_{n-k}\,.$$

(b) In (2.61) (a) it is actually sufficient to have that $(*)$ commutes on models M after composition with ι_2^M, similar in (b) and the lemma below. It seems that ∇ may have further components of negative lower degree, but this is not needed.

(c) An instance of (2.61) (a) is proved similarly in [BouG].

Lemma (2.64)

Let n be a fixed integer $\geqslant 2$. Let $B_n : \mathcal{T} \to \langle \mathrm{DG}^*\rangle$ and $A : \mathcal{T} \to \langle \mathrm{dg}^*\rangle$ be cofunctors. Suppose A is representable with models $\mathcal{M}$ and $B_n(M)$ acyclic rel V^M concentrated in degree $-n$. Let $\eta : K \rightleftarrows A : \varepsilon$ be natural chain maps with $\varepsilon\eta = 1$. Let $g : B_n \to A$ be a natural map of degree 2 with $Dg = 0$ and $\varepsilon g = 0$. In case $n = 2$ assume that $g^{-2} = 0$.

Then there exists $f : B_n \to A$ with $Df = g$ and $\varepsilon f = 0$. □

This lemma is straightforward using the method of acyclic models, i.e. the dual of (2.39). First construct f' with $Df' = g$, then put $f = (1 - \eta\varepsilon)f'$.

Proof of (2.61). *Ad* (a). It suffices to show there exist graded maps $v_n : B_n \to A$ $(n \geqslant 2)$ such that

$$(*)_n \qquad v_n d_n + dv_n = -v_{n-1}\delta_n + \textstyle\sum_{k=1}^{n-1} v_k \cup v_{n-k}$$

and $\varepsilon v_n = 0$. Notice that by assumption v_1 has these properties (put $v_0 = 0$). Then $v = \sum v_n : B \to A$ is the desired twisting map. In fact, since B is Σ-decomposable, $(*)_n$ says that on B_n we have $Dv = vd + v\delta + dv = v \cup v$.

By induction on $n \geqslant 2$, suppose we have v_p with $(*)_p$ and $\varepsilon v_p = 0$, for all $p < n$. We have to construct v_n satisfying $D^{\mathrm{int}}v_n = g$ where g is the right-hand side of

$(*)_n$ and $D^{\rm int}$ is the differential of $\mathrm{Hom}(B_n, A)$ where B_n is regarded as a chain complex with the internal differential d. By (2.64) it suffices to show $D^{\rm int}g = 0$ and $\varepsilon g = 0$. The latter is trivial. It is convenient to put $x = \sum_{k=0}^{n-1} v_k$. Then $x_k = v_k$ and (i) $(x \cup x)_k = \sum_{\ell=1}^{k-1} v_\ell \cup v_{k-\ell}$ for $k \leqslant n$, (ii) $D(x)_k = (x \cup x)_k$ for $k < n$ by the inductive hypothesis.

We have $D^{\rm int}(v_{n-1}\delta_n) = dv_{n-1}\delta_n - v_{n-1}\delta_n d_n = -dv_{n-1}\delta_n + v_{n-1}d_{n-1}\delta_n = (x \cup x)_{n-1}\delta_n$ where we use $(*)_{n-1}$ and (i). The following computation completes the proof:

$$\begin{aligned}
D^{\rm int}g &= D^{\rm int}[(x \cup x)_n - v_{n-1}\delta_n] \\
&= d(x \cup x)_n - (x \cup x)_n d_n - (x \cup x)_{n-1}\delta_n \\
&= [d(x \cup x) - (x \cup x)d - (x \cup x)\delta]_n \\
&= [D(x \cup x)]_n \\
&= [D(x) \cup x - x \cup D(x)]_n = \textstyle\sum_{k=1}^{n-1} D(x)_k \cup x_{n-k} - x_k \cup D(x)_{n-k} \\
&= \textstyle\sum_{k=1}^{n-1} (x \cup x)_k \cup x_{n-k} - x_k \cup (x \cup x)_{n-k} \\
&= [(x \cup x) \cup x - x \cup (x \cup x)]_n \\
&= 0\,.
\end{aligned}$$

Ad (b). This is proved by the same method. Put $h_0 = \mathbb{1}$ and construct inductively $h_n : B_n \to A$ $(n \geqslant 1)$ such that

$$(**)_n \qquad h_n d_n - dh_n = -h_{n-1}\delta_n + v_n - w_n + \textstyle\sum_{k=1}^{n-1} v_k \cup h_{n-k} - h_k \cup w_{n-k}$$

and $\varepsilon h_n = 0$. The existence of h_n is guaranteed by (2.64) where a computation similar to (a) is to be done. The details are omitted. □

Our reflections on duality in N° 3 are in part motivated by the question whether it is possible to obtain sort of a dual result without writing down a dual proof. The answer is yes, but the argument is not quite obvious.

Dualization (2.65)

There is an analogous result for covariant functors. Then $B : \mathcal{T} \to \langle \mathrm{DGA} \rangle_0$ is to be a Π-*decomposable* functor in the sense that $B = (\prod_{n=0}^{\infty} B^n, \nabla = d + \partial)$ where B_*^* is a bigraded algebra with differentials d, ∂ of upper degree 0 resp. 1 (as in the cobar construction). We get a result on the existence of twisting maps and homotopies $C \to B$ provided $C : \mathcal{T} \to \langle \mathrm{dgC} \rangle_0$ is representable and B^n acyclic on models rel "something in degree $-n$".—In order to prove it, one may argue as follows. By a limit argument one is reduced to the case where $B^n = 0$ for n large. Since C is positive we may also assume that $B^n_{<-2n} = 0$ for all n. Then B is bounded below and the result follows from (2.61), since categorical duality is a monoidal co-equivalence on bounded below objects. Details are left to the reader.

The next result is formulated for covariant functors. Here it clear that categorical dualization yields an equivalent result. An instance of (a) is proved similarly in [May].

Theorem (2.66)

Let $C : \mathcal{T} \to \langle \mathrm{dgC} \rangle_1$ *and* $A : \mathcal{T} \to \langle \mathrm{dgA} \rangle_0$ *be functors. Suppose* C *is degreewise representable with models* $\mathcal{M}$. *Suppose* $A(M)$ *is acyclic rel* $\rho^M : A(M) \to V^M$ *with* V^M *concentrated in degree* 0 $(M \in \mathcal{M})$.

(a) *Given* $v_1 : C_1 \to A_0$, $v_2 : C_2 \to A_1$ *with* $\varepsilon v_1 = 0$ *and* $dv_2 + v_1 d = v_1 \cup v_1$, *there is a twisting map* $v : C \to A$ *extending* v_1, v_2.

(b) *Given twisting maps* $v, w : C \rightrightarrows A$ *with* $v_1 = w_1$, *there is a twisting homotopy* $h : C \to A$, $v \simeq w$. [Actually it is sufficient that $\rho^M(v_1 - w_1)_M = 0$.]

Proof. *Ad* (a). Put $v_{\leqslant 0} = 0$. We have to construct $v_3, v_4, \ldots$ such that (for all n)

$$(*)_n \qquad d\,v_n = -v_{n-1} d + \textstyle\sum_{k=1}^{n-1} v_k \cup v_{n-k} .$$

Then $v = \sum v_n$ is a twisting map, because $(*)_n$ is the degree n part of the equation $dv = -vd + v \cup v$. — By induction on $n > 2$, suppose we have v_p satisfying $(*)_p$, for $p < n$. Let $y : C_n \to A_{n-2}$ be the right-hand side of $(*)_n$. A straightforward computation yields $d\,y = 0$ (similar to the proof of the preceding result, but easier). Moreover, $\rho^M y_M = 0$ for reasons of degree. Hence by (2.39) there is v_n with $d\,v_n = y$. This completes the inductive construction of v.

Ad (b). Put $h_{<0} = 0$, $h_0 = \mathbb{1}$. We have to construct $h_1, h_2, \ldots$ such that (for all n)

$$(**)_n \qquad dh_n = h_{n-1} d + v_n - w_n + \textstyle\sum_{k=1}^{n-1} v_k \cup h_{n-k} - h_k \cup w_{n-k} .$$

This is done by induction using (2.39) as in the proof of (a). □

§4 EZ-Morphisms

There is a general theorem in [GuM] showing that $\mathcal{AW} : C(X \times Y) \to C(X) \otimes C(Y)$ is strongly homotopy comultiplicative, i.e. can be extended to a map between the corresponding cobar constructions. We recall and extend this theorem in N°1. In N°2 we prove a generalization which is some kind of "extension lemma" for twisting maps. The method is similar to that of acyclic models, but here the constructions are unique. Some new results are then obtained in N°3.

N°1 Extension of an EZ-morphism

We use ${}^{\xi}C$ or C^{ξ} to denote a chain complex together with a map $\xi : C \to C$ of degree $+1$.

An *EZ-morphism* is a diagram

$$(2.67) \qquad {}^{\xi}C \underset{\varphi}{\overset{\nabla}{\leftrightarrows}} T$$

where $\nabla \in \langle \mathrm{dgC} \rangle_0$, φ a chain map and ξ a chain homotopy: $\nabla\varphi \simeq 1_C$ (i.e. $D\xi = \nabla\varphi - 1_C$) such that $\varphi\nabla = 1_T$, $\xi\nabla = 0$, $\xi\xi = 0$, $\varphi\xi = 0$.

It follows that $\xi\eta = 0 = \varepsilon\xi$, and that φ preserves counit ε and coaugmentation η. The definition makes sense also in the non-augmented case where $\langle \mathrm{dgC} \rangle_0$ is replaced by $\langle \mathrm{dgC} \rangle$. For the main source of examples see App. B, N°2.

Let $(T, \nabla, \varphi, C, \xi)$ *be a fixed EZ-morphism where* C, $T \in \langle \text{dgC} \rangle_2$.

In essence the following theorem says (a) there is a distinguished twisting map f as in the diagram below, which in a sense 'extends' φ (notice that $t_T\varphi$ need not be a twisting map) and (b) there are distinguished twisting homotopies $h : \Omega(\nabla)f \simeq t_C : \bar{h}$.

Recall that ΩC is a tensor algebra except for the differential. Since here C is 2-reduced we have $\Omega C = \prod_{k=0}^{\infty} (s_{-1}\widetilde{C})^{\otimes k}$, i.e. the direct sum (being finite in each degree) *is* a product.

THEOREM (2.68)

(a) *There is unique twisting map* f *satisfying* $f\nabla = t_T$, $f\xi = 0$ *and* $f^1 = t_T\varphi$.

(b) *There is unique twisting homotopy* $h : \Omega(\nabla)f \simeq t_C$ *satisfying* $h\nabla = \mathbb{1}$, $h\xi = 0$ *and* $h^1 = -t_C\xi$. *Further* $Fh = \mathbb{1}$. — *There is also unique twisting homotopy* $\bar{h} : t_C \simeq \Omega(\nabla)f$ *satisfying* $\bar{h}\nabla = \mathbb{1}$, $\bar{h}\xi = 0$ *and* $\bar{h}^1 = t_C\xi$. *Further* $F\bar{h} = \mathbb{1}$.

(c) *These maps* f, h, $\bar{h}$ *are natural with respect to transformations of EZ-morphisms.*

The proof is given in N° 2. Notice that by the universal properties of the cobar construction we get a diagram

$$^{H}\,\Omega C \underset{F}{\overset{\Omega\nabla}{\rightleftarrows}} \Omega T$$

where F is the map in $\langle \text{dgA} \rangle_1$ induced by f and satisfying $F \circ \Omega\nabla = 1$, while $H : \Omega\nabla \circ F \simeq 1$ is the derivation homotopy induced by h. Similarly we have $\mathcal{H} : 1 \simeq \Omega\nabla \circ F$ induced by $\bar{h}$. We call f or F the *extension* of φ (for the given EZ-morphism). Similarly with the homotopies.

The theorem has a dual which holds without any reducedness assumption, see N° 2. Except for the uniqueness parts and the general context — and an apparently incorrect formula in their (apart from that omitted) proof of (b) — the theorem is established in [GMu].

N° 2 A generalization

Here one can work with $\mathbb{Z}$-graded (co)algebras, but we refrain from this generality. Note that if A^*_* is a bigraded algebra then $A = \prod_{k=0}^{\infty} A^k$ is a graded algebra where the n-th component of the multiplication is the obvious composite $A \otimes A \to \bigoplus_{k=0}^{n} A^k \otimes A^{n-k} \to A^n$. We say that $A \in \langle \text{dgA} \rangle$ is Π-*decomposable* if $A = \prod_{k=0}^{\infty} A^k$ where A^*_* is a bigraded algebra with $A^0 = K$. Note that A has a canonical augmentation. (Here we need no condition on the differential of A).

Let $(T, \nabla, \varphi, C, \xi)$ be a fixed EZ-morphism where C, $T \in \langle \text{dgC} \rangle_0$. In addition, let $A \in \langle \text{dgA} \rangle_0$ be Π-decomposable and let u be a twisting map as in ...

(2.69)

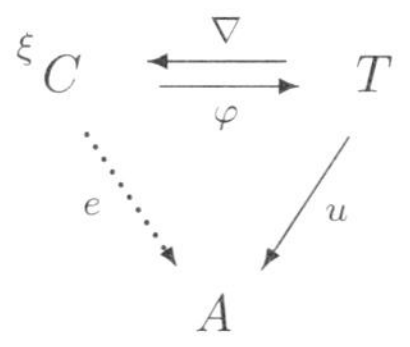

Theorem (2.70)

(a) *There is a twisting map* $e : C \to A$ *satisfying* $e\nabla = u$, $e\xi = 0$, $e^1 = u^1\varphi$ *and it is unique: Two twisting maps* e, f *satisfying* $e\nabla = f\nabla (= u)$ *and* $e\xi = f\xi$ *coincide.*

(b) *Let* $e, f : C \rightrightarrows A$ *be twisting maps where* $e\nabla = f\nabla (= u)$. *There is unique twisting homotopy* $h : e \simeq f$ *resp.* $\bar{h} : f \simeq e$ *satisfying* $h\nabla = \mathbb{1}$, $h\xi = 0$, $h^1 = (e^1 - f^1)\xi$ *resp.* $\bar{h}\nabla = \mathbb{1}$, $\bar{h}\xi = 0$, $\bar{h}^1 = (f^1 - e^1)\xi$.

Proof. *Ad* (a). We define $e^n : C \to A^n$ (also regarded as a map into A) inductively as follows:

$$\begin{aligned} e^0 &= 0 \\ e^n &= -\textstyle\sum_{k=0}^{n}(e^k \cup e^{n-k})\xi + u^n\varphi \end{aligned}$$

Then we get $e = \sum_n e^n : C \to A$. Since A^*_* is a bigraded algebra it follows that e satisfies

$$(*) \qquad e = -(e \cup e)\xi + u\varphi\,.$$

From this relation it is obvious that e has the stated properties. Also $e\eta = 0$, $\varepsilon e = 0$ are obvious and $De = e \cup e$ is seen as follows. First $(*)$ implies without difficulty

$$De = (-De \cup e + e \cup De)\xi + e \cup e$$

or $x = (-x \cup e + e \cup x)\xi$ where $x = De - e \cup e$. Thus

$$x^n = \textstyle\sum_{k=0}^{n}(-x^k \cup e^{n-k} + e^k \cup x^{n-k})\xi$$

which shows $x^n = 0$ by induction. Hence $x = 0$ and e is a twisting map. The uniqueness is shown similarly, one has to check that every twisting map e with the stated properties satisfies $(*)$. For the strong uniqueness assertion, see end of proof.

Ad (b). Define $h = \sum h^n$ where

$$\begin{aligned} &h^0 = \mathbb{1}\,, \qquad h^1 = (e^1 - f^1)\xi\,, \\ &h^n = \textstyle\sum_{k=0}^{n}(e^k \cup h^{n-k} - h^k \cup f^{n-k})\xi\,, \quad \text{for } n \geqslant 1\,. \end{aligned}$$

Then clearly

$$(**) \qquad h = (e \cup h - h \cup f)\xi + \mathbb{1}\,.$$

Put $z = e \cup h - h \cup f$. We have to show that $Dh - z = y$ is zero. We first claim: $y = -(e \cup y + y \cup f)\xi$. Indeed,

$$[e \cup (Dh - z) + (Dh - z) \cup f]\xi$$

$$\begin{aligned} &= [e \cup Dh - e \cup e \cup h + \{e \cup h \cup f\} + Dh \cup f - \{e \cup h \cup f\} + h \cup f \cup f]\xi \\ &= -(Dz)\xi \end{aligned}$$

$$\begin{aligned} &\overset{(**)}{=} && -(Dh + z(D\xi)) \\ &= && -(Dh - z)\,. \end{aligned}$$

The last equation is verified by

$$\begin{aligned} z(D\xi) + z &= z\nabla\varphi \\ &= (e \cup h - h \cup f)\nabla\varphi \\ &= (e\nabla \cup h\nabla - h\nabla \cup f\nabla)\varphi \\ &= (u \cup 1\!\!1 - 1\!\!1 \cup u)\varphi \\ &= 0 \end{aligned}$$

where we use that ∇ is comultiplicative and the obvious relation $h\nabla = 1\!\!1$. This proves our claim. We conclude that

$$y^n = -\sum_{k=0}^{n}(e^k \cup y^{n-k} + y^k \cup f^{n-k})\xi$$

and thus $y^n = 0$ by induction. Hence $y = 0$ as desired. The uniqueness of h follows from the fact that any $h : e \simeq f$ with $h\nabla = 1\!\!1$, $h\xi = 0$ satisfies $(**)$.

Finally we prove the strong uniqueness assertion in (a). If $(e - f)\xi = 0$ then by definition $h = 1\!\!1$, thus $0 = Dh = e \cup h - h \cup f = e - f$. □

Remarks (2.71)

(a) It is clear that these constructions are natural in the sense that (2.69) may be regarded as a diagram of functors and natural maps. The decomposition of A however need not be natural.

(b) The existence of a twisting map e with $e\nabla = u$ and $e\xi = 0$ follows from the case $A = \Omega T$ and $u = t_T$ by the universal property of t_T. Hence for this we do not need that A is decomposable provided the extension of φ exists.

(c) The theorem above implies a dual result, cf. § 3, N° 3. Hence the dual of (2.68) holds without the assumption of 2-reducedness. The point is that the bar construction has a Σ decomposition by definition, while the cobar construction has a Π-decompostion only under that assumption.

(d) The assumption on A implies that this is a complete algebra (for the augmental filtration). This seems sufficient here. Instead one may try with the condition that C is c-connected such that ξ sends F_nC into $F_{n+c}C$; cf. (2.13).

Proof of (2.68). We only have to verify $Fh = 1\!\!1$. Since Fh and $1\!\!1$ are twisting homotopies from $F\Omega(\nabla)f = f$ to $Ft_C = f$ satisfying (i) $(Fh)\nabla = F1\!\!1 = 1\!\!1 = 1\!\!1\nabla$, (ii) $(Fh)\xi = F0 = 0 = 1\!\!1\xi$, the assertion follows from uniqueness (2.70) (b). □

N° 3 Properties of the extension

We establish further properties of the constructions provided by (2.68).

Observe first that we can tensorize our EZ-morphism with some $B \in \langle \mathrm{dgC} \rangle_2$ obtaining new EZ-morphisms:

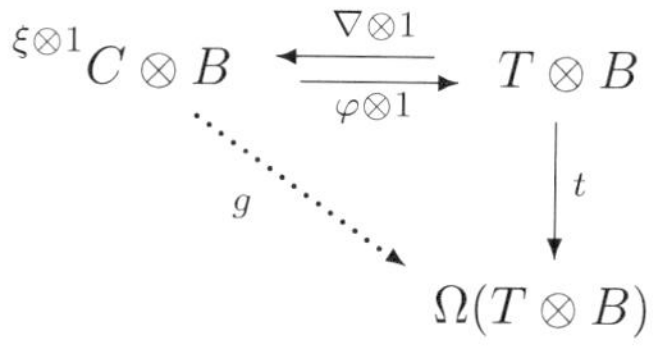

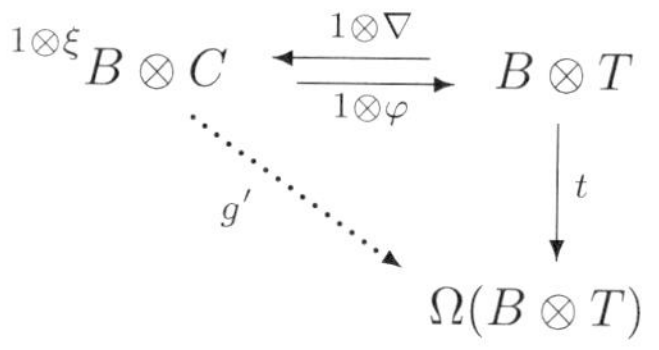

By virtue of (2.68) we obtain the well-defined extension (twisting map) g resp. g'. We denote by G resp. G' the corresponding dgA-map.

LEMMA (2.72)

The following diagram commutes.

$$C \otimes B \xrightarrow{t} \Omega(C \otimes B) \xrightarrow{G} \Omega(T \otimes B)$$
$$t*t \searrow \quad j \downarrow \quad \quad j \downarrow$$
$$\Omega C \otimes \Omega B \xrightarrow{F \otimes 1} \Omega T \otimes \Omega B$$

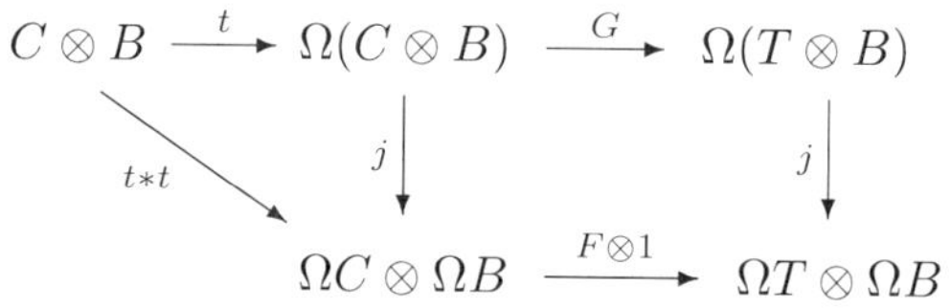

There is an analogous result relating G' to $1 \otimes F$.

Proof. The left triangle commutes by definition of j. Thus we have to show that $jg = f*t$. Of course $\Omega T \otimes \Omega B$ has a canonical Π-decomposition. Hence by (2.70)(a) we only have to compute: $(jg)(\nabla \otimes 1) = j t_{T \otimes B} = t_t * t_B = (f * t_B)(\nabla \otimes 1)$, $(jg)(\xi \otimes 1) = 0 = 0 * 0 = f\xi * 0 = (f * t_B)(\xi \otimes 1)$. $\square$

Next let there be given another EZ-morphism $(C, \nu, \psi, B, \kappa)$ where B is 2-reduced. We denote the extension of ψ by g. Regard the diagram

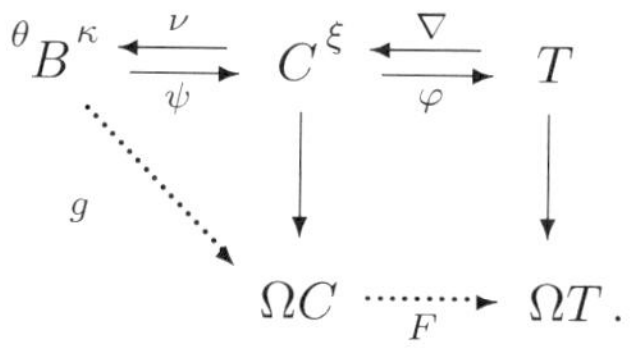

One checks that $(T, \nu\nabla, \varphi\psi, B, \theta)$ with $\theta = \kappa + \nu\xi\psi$ is an EZ-morphism, the *composite* of the given ones.

LEMMA (2.73)

Fg is the extension of $\varphi\psi$ for the composite EZ-morphism.

Proof. By (2.70)(a) this follows from $Fg(\nu\nabla) = Ft_C\nabla = f\nabla = t_T = e(\nu\nabla)$ and $Fg\theta = Fg(\kappa + \nu\xi\psi) = Fg\nu\xi\psi = Ft_C\xi\psi = f\xi\psi = 0 = e\theta$. $\square$

Remark. There is an obvious identity EZ-morphism for this composition. Recall that we can form the tensor product of an EZ-morphism with the identity (from left or right). Using composition one may define (two) tensor products of EZ-morphisms in general.

The third and final lemma is straightforward by naturality of the twisting homotopy relation. Let $(T, \nabla, \varphi, C, \xi)$ and $(T, \nabla, \varphi, C, \xi')$ be EZ-morphisms (only their chain homotopies are different). Let f, f' be the corresponding extensions of φ.

LEMMA (2.74)
The map $F'h : C \to \Omega T$ is a twisting homotopy $F'h : f \simeq f'$. □

Appendix B CHAIN (CO)FUNCTORS

N° 1 Monoidal categories

Recall ([Mc], [Mc3]) that a (symmetric, as always) *monoidal category* is a category $\mathcal{V}$ together with a functor $\otimes : \mathcal{V} \times \mathcal{V} \to \mathcal{V}$, a *unit* object K in $\mathcal{V}$ and natural isomorphisms $\alpha_{A,B,C} : A \otimes (B \otimes C) \cong (A \otimes B) \otimes C$, $\lambda_A : K \otimes A \cong A$, $\rho_A : A \otimes K \cong A$, $\tau_{A,B} : A \otimes B \cong B \otimes A$. The natural isomorphism $\alpha, \lambda, \rho, \tau$ are assumed to satisfy the condition of coherence.—Roughly speaking, *coherence* means that from one 'permuted iterate' of the functor $\otimes$ to another such there is a *unique* natural isomorphism built up by 'instances' of these four, there inverses and the identity. This coherence follows from the commutativity of only a small number of diagrams: a pentagon involving α, a hexagon involving α and τ, plus the equalities $\lambda_K = \rho_K$, $\rho_A = \lambda_A \circ \tau_{A,K}$, $\tau_{B,A} \circ \tau_{A,B} = 1_{A\otimes B}$ and $(\rho_A \otimes 1_C) \circ \alpha_{A,K,C} = 1_A \otimes \lambda_C$.

(Co)monoids (B.1)

Let $\mathcal{V}$ be a monoidal category. There is a notion of *monoid* (= unital, associative monoid) and of *commutative* monoid. There is a dual notion of *comonoid* etc. (For more details in a special situation, see Chp. 1, § 3, N° 1.) Coherence implies a general (commutativity and) associativity law for the multiplication of a (commutative) monoid; dually for comonoids. Moreover K is a (co)commutative (co)monoid in a canonical way.

We denote by Mon $\mathcal{V}$, cMon $\mathcal{V}$, Com $\mathcal{V}$, and cCom $\mathcal{V}$ the categories of monoids, commutative monoids, comonoids, and cocommutative comonoids, respectively (the morphisms are defined in the obvious manner). In the two former categories, K is an initial object, in the two latter K is terminal. Hence we obtain *pointed* categories (i.e. there is a common initial and terminal object) $\mathrm{Mon}_0\,\mathcal{V}$ etc. An object of $\mathrm{Mon}_0\,\mathcal{V}$, called *based monoid*, is a pair (A, ε) where $\varepsilon : A \to K \in \mathrm{Mon}\,\mathcal{V}$, and a morphism is a map in Mon $\mathcal{V}$ commuting with ε. (The other cases are analogous.)

These categories are again monoidal. In fact, if A, B are (based) monoids so is $A \otimes B$ with multiplication $(m_A \otimes m_B)\tau_{2,3}$ (where $\tau_{2,3} : S\otimes T\otimes U\otimes V \to S\otimes U\otimes T\otimes V$ is the *canonical* isomorphism, i.e. $\tau_{2,3} = 1 \otimes \tau \otimes 1$). A monoid is commutative iff its multiplication is a morphism of monoids.

Functors and monoidal transformations (B.2)

Let $\mathcal{T}$ and $\mathcal{V}$ be monoidal categories and $F : \mathcal{T} \to \mathcal{V}$ a functor. There are several conditions expressing to which extend F is "compatible with tensor products". A *monoidal transformation* for F consists of a natural map $\phi : F(X) \otimes F(Y) \to F(X \otimes Y)$ and a map $u : K \to F(K')$, such that ϕ is associative and unital with u. (Dually one may have maps in the other direction.) Here ϕ (with u) is *unital* if $F(\lambda_X)\phi_{K,X}(u \otimes F(X)) = \lambda_{F(X)}$ and analogously with ρ, — *associative* if $\phi(1 \otimes \phi)\alpha = F(\alpha)\phi(\phi \otimes 1)$, — *commutative* if $F(\tau)\phi = \phi\tau$.

When F is equipped with a monoidal transformation, there is an induced functor between the respective categories of monoids, i.e. $F : \mathrm{Mon}\,\mathcal{T} \to \mathrm{Mon}\,\mathcal{V}$. When u is an isomorphism (resp. ϕ commutative), the same holds with based (resp. commutative) monoids. A functor F with a commutative monoidal transformation ϕ, u consisting of isomorphisms is said to be a *monoidal functor*.

Now let $F, G : \mathcal{T} \rightrightarrows \mathcal{V}$ be functors with monoidal transformations ϕ, u and ψ, v (in the same direction). Let $f : F \to G$ be a natural map satisfying $\psi(f_X \otimes f_Y) = f_{X\otimes Y}\phi$ and $f_{K'}u = v$ (in other words f is "compatible with tensor products"). Then f yields a natural map between the induced functors $F, G : \mathrm{Mon}\,\mathcal{T} \rightrightarrows \mathrm{Mon}\,\mathcal{V}$ etc.

Additive monoidal category (B.3)

We call $\mathcal{V}$ *additive monoidal*, if $\mathcal{V}$ is an additive category and also a (symmetric, as always) monoidal category, such that $\otimes$ is additive in each variable. We often have to assume that $\mathcal{V}$ is *strongly additive monoidal*, i.e. $\mathcal{V}$ has arbitrary direct sums and these are preserved by $\otimes$ in each variable.

In this setting we call (based) monoids also (*augmented*) *algebras* — and (based) comonoids also (*coaugmented*) *coalgebras* — over $\mathcal{V}$. We call the cocommutative comonoids over $\langle\mathrm{Alg}\rangle_{\mathcal{V}}$ also *Hopf algebras* — and the commutative monoids over $\langle\mathrm{Coalg}\rangle_{\mathcal{V}}$ also *Hopf coalgebras* — over $\mathcal{V}$.

Recall that $\langle\mathrm{DG}\rangle_{\mathcal{V}}$ and $\langle\mathrm{s}\rangle_{\mathcal{V}} = {}^{\Delta}\mathcal{V}$ receive additive monoidal structure. In these cases the algebras are called DG algebras resp. simplicial algebras, etc.

Lie algebra theory? (B.4)

Also a category of *Lie algebras* over $\mathcal{V}$ may be defined in this setting. In fact, the anti-commutativity and the Jacobi identity can be expressed using canonical interchange isomorphisms. Note however that the classical definition of Lie algebra over K includes the condition $[x, x] = 0$, i.e. that "squares are zero" (when 2 is invertible in K this follows from the anti-commutativity). This condition is hard to express in the general setting. However this could be eluded by assuming that Lie algebras can be embedded into an associative algebra. Anyway we do not try to include in the present chapter any Lie (or commutative) algebra theory. Hence most of this chapter does not depend on the assumption that the tensor product is symmetric. Compare the discussion in Chp. 3, App. C, N° 4.

N° 2 Normalization

Let $\mathcal{V}$ be an additive category with (co)kernels.

Normalization (B.5)

The *Dold - Kan correspondence* says that the category of simplicial objects over $\mathcal{V}$ is equivalent to the category of chain complexes over $\mathcal{V}$. This equivalence is given by the *normalization* functor

(B.6) $$N : \langle\mathrm{s}\rangle_{\mathcal{V}} \xrightarrow{\sim} \langle\mathrm{dg}\rangle_{\mathcal{V}}$$

where $N(A)$ is defined by $(NA)_n = \ker[(d_i) : A_n \to \prod_{i=1}^{n} A_{n-1}]$, the "intersection" of the kernels of the d_i ($i \geqslant 1$), endowed with $d = d_0$ as differential.

Regard the simplicial object A as a chain complex with the standard differential $d = \sum_{i=0}^{n+1}(-1)^i d_i : A_n \to A_{n-1}$. We have the natural decomposition of chain subcomplexes $A = N(A) \oplus D(A)$ where $D(A)_n = \operatorname{im}[(s_i) : \bigoplus_{i=0}^{n-1} A_{n-1} \to A_n]$ is the graded subobject "generated by" the images of the degeneracy operators. The *Normalization Theorem* of Eilenberg - Mac Lane says that $D(A)$ is contractible, there is a natural chain homotopy $1 \simeq 0$. Hence the canonical (retraction) chain maps $A \rightleftarrows N(A)$ constitute a chain homotopy equivalence.

Because of the canonical isomorphism $NA \cong A/D(A)$ we may regard NA either as a sub- or as a quotient complex of A.

For sufficiently general arguments see [DPu], [Ep].

Compatibility with tensor products (B.7)

Suppose that $\mathcal{V}$ is strongly additive monoidal with unit object K. The categories in (B.6) have canonical symmetric monoidal structure. The (degreewise) tensor product in ${}^{\Delta}\mathcal{V} = \langle \mathrm{s} \rangle_{\mathcal{V}}$ is denoted by $\otimes$; and the unit object in both cases by K. We have a canonical isomorphism $N(K) = K$; and the *Eilenberg - Zilber* theorem says that there are natural chain maps

(B.8) $$\mathcal{AW} : N(A \otimes B) \rightleftarrows N(A) \otimes N(B) : \mathcal{EM}$$

with $\mathcal{AW} \circ \mathcal{EM} = 1$ and there is a natural chain homotopy $h : \mathcal{EM} \circ \mathcal{AW} \simeq 1$. One can choose h such that $h \circ \mathcal{EM} = 0$, $h \circ h = 0$, $\mathcal{AW} \circ h = 0$ ([EMc]). We use the following explicit formulas.

The *Alexander - Whitney* map.—Consider the map $A \otimes B \to A \otimes B$ which in degree n is defined as

$$\mathcal{AW} = \sum_{i=0}^{n} \tilde{d}^{n-i} \otimes d_0^i \, .$$

Here $\tilde{d}$ denotes the boundary operator with highest index. One checks that this chain map induces a (unique) map $\mathcal{AW}$ as in (B.8) where $N(A)$ etc. is regarded as a quotient of A.

The *Eilenberg - Mac Lane* map. Consider the map $A \otimes B \to A \otimes B$ which in degree n and on the summand $A_p \otimes B_{n-p}$ is defined as

$$\mathcal{EM} = \sum_{\sigma} (-1)^{\varepsilon(\sigma)} s_{\sigma(n-1)} \cdots s_{\sigma(p)} \otimes s_{\sigma(p-1)} \cdots s_{\sigma(0)} \, .$$

The sum is taken over all permutations of $\{0, \ldots, n\}$ which preserve order on the first p and on the last $n-p$ elements. As above this chain map induces the desired map $\mathcal{EM}$.

It is well-known (and verified by computation) that together with the isomorphism $N(K) = K$ these are monoidal transformations (i.e. associative and unital). Moreover $\mathcal{EM}$ is commutative, but $\mathcal{AW}$ is *not*. For terminology, see (B.2).

Hence normalization induces functors, still denoted by N,

(B.9) $$\boldsymbol{N} : \begin{cases} \langle \mathrm{sCC} \rangle \longrightarrow \langle \mathrm{dgC} \rangle_0 \, , & \langle \mathrm{sCoalg} \rangle \longrightarrow \langle \mathrm{dgC} \rangle \\ \langle \mathrm{sAA} \rangle \longrightarrow \langle \mathrm{dgA} \rangle_0 \, , & \langle \mathrm{sAlg} \rangle \longrightarrow \langle \mathrm{dgA} \rangle \\ \langle \mathrm{sLA} \rangle \longrightarrow \langle \mathrm{dgL} \rangle \, . & \end{cases}$$

For the latter we need the commutativity of $\mathcal{EM}$ which also implies that the property of being commutative is preserved by N in the case of algebras.

Perhaps not so familiar though of great importance is the following. Recall that the tensor product gives symmetric monoidal structure in all the above categories of (co)algebras.

Proposition (B.10)
In case A, $B \in \langle \text{sAA} \rangle$ *the maps* $\mathcal{EM}$, $\mathcal{AW}$ *are in* $\langle \text{dgA} \rangle_0$.
In case A, $B \in \langle \text{sCC} \rangle$ *the map* $\mathcal{EM}$ *is in* $\langle \text{dgC} \rangle_0$ (but $\mathcal{AW}$ is *not*).

Hence in the second case diagram (B.8) together with h is an EZ-morphism in $\langle \text{dgC} \rangle_0$. There is an analogous result in the non-augmented case.

Proof. From these three assertions the first follows from the mentioned properties of $\mathcal{EM}$. The other two follow from a certain distributivity relation between $\mathcal{EM}$ and $\mathcal{AW}$, cf. [GMay | App.]. □

By construction the quotient chain map $A \to NA$ is compatible with the monoidal transformations $\mathcal{EM}$ and $\mathcal{AW}$ (recall these are also defined using the non-normalized chain complex). The inclusion $NA \to A$ is compatible with $\mathcal{EM}$ but *not* with $\mathcal{AW}$.

Chain functors (B.11)
We define the *chain functor* $K : \mathcal{S} \to \langle \text{s} \rangle_{\mathcal{V}}$ by $X \mapsto KX := X \otimes K$. The *normalized chain functor* is $C(X) = C(X; K) := N(KX)$. Notice that $C(X)$, viewed as a quotient of the chain complex KX, is the K-free graded object on the non-degenerate simplices of X. These definitions make sense with K replaced by any (coefficient) object $W \in \mathcal{V}$; monoidal structure is not needed. But if as above K is the unit object for a tensor product on $\mathcal{V}$ then we clearly have $C(X; W) = C(X; K) \otimes W$.

The chain functor is clearly monoidal, hence it gives rise to functors $K : \mathcal{S} \to \langle \text{sCoalg} \rangle$, $\mathcal{S}_0 \to \langle \text{sCC} \rangle$, resp. $\langle \text{sGp} \rangle \to \langle \text{sAA} \rangle$. Composition with the functors in (B.9) yields functors, all denoted by C,

$$\text{(B.12)} \qquad \boldsymbol{C} : \begin{cases} \mathcal{S}_0 \longrightarrow \langle \text{dgC} \rangle_0 , \qquad \mathcal{S} \longrightarrow \langle \text{dgC} \rangle \\ \qquad\qquad \langle \text{sGp} \rangle \longrightarrow \langle \text{dgA} \rangle_0 . \end{cases}$$

As noted earlier ((2.38), (2.43)) the normalized chain functor $C : \mathcal{S} \to \langle \text{dgC} \rangle$ is degreewise representable and acyclic on models rel $\varepsilon : C \to K$. We call any functor $F : \mathcal{S} \to \langle \text{dgC} \rangle$ with these properties a *homology theory*. In fact, given such F there is a natural chain map $f : C \to F$, and moreover f is a chain homotopy equivalence commuting with the diagonals up to chain homotopy. This is proved by the method of acyclic models, for more details in the dual situation, see N° 4.

N° 3 Representable cofunctors for spaces

We prefer here the contravariant setting, since this is more important for our applications, moreover in this way we avoid *co*simplicial objects. We assume that $\mathcal{V}$ has arbitrary limits. Our main objective is to show that representable cofunctors $\mathcal{S} \to \mathcal{V}$ are in natural one-one correspondence with contractible simplicial objects over $\mathcal{V}$. This is a means for constructing such cofunctors.

The normalized cochain functor (B.13)

Dualizing the construction of the normalized chain functor (B.11), we obtain the *normalized cochain functor* $C^*(-;W) : \mathcal{S} \to \langle \mathrm{dg}^* \rangle_{\mathcal{V}}$ where $W \in \mathcal{V}$ is any (coefficient) object. Fortunately there is another description, so that the dualization of the constructions in N° 2 can be avoided. In fact, it makes sense to define $C^*(X;W) = \mathrm{Hom}(C(X;\mathbb{Z}), W)$ where $\mathbb{Z}$ is the ring of integers.

Here we use the "external" Hom-functor

$$\mathrm{Hom} : (\mathcal{M}od_{\mathbb{Z}}^{\mathrm{free}})^{\mathrm{op}} \times \mathcal{V} \longrightarrow \mathcal{V}$$

which is essentially well-defined by the conditions: $\mathrm{Hom}(-,V)$ carries direct sums to products and $\mathrm{Hom}(\mathbb{Z},-) = \mathrm{Id}$. The construction of this functor is straightforward; it can be extended (essentially uniquely) to $(\mathcal{M}od_{\mathbb{Z}})^{\mathrm{op}} \times \mathcal{V}$, such that $\mathrm{Hom}(-,V)$ carries colimits to limits. This however is not needed. We have the following natural isomorphism (adjunction)

$$\mathcal{V}(W, \mathrm{Hom}(A,V)) \approx \mathcal{M}od_{\mathbb{Z}}(A, \mathcal{V}(W,V))$$

from which the uniqueness and limit properties follow. For the perfectly dual construction of $\otimes : \mathcal{M}od_{\mathbb{Z}} \times \mathcal{V} \to \mathcal{V}$, cf. [DPu|3.32].

Restriction and extension (B.14)

Any cofunctor $F : \mathcal{S} \to \mathcal{V}$ has a *restriction* $|F| = F \circ \Delta[-] \in {}^{\Delta}\mathcal{V}$ where $\Delta[-] : \Delta \to \mathcal{S}$ is the standard (full) embedding; analogously for natural maps resp. simplicial maps. It is convenient to use the cofunctor category ${}^{\mathcal{S}}\mathcal{V} = \mathcal{V}^{\mathcal{S}^{\mathrm{op}}}$. Then we have the restriction functor

$$\text{restriction } |-| : {}^{\mathcal{S}}\mathcal{V} \rightleftarrows {}^{\Delta}\mathcal{V} : -^* \text{ extension}.$$

There is also a canonical *extension functor* which is constructed next. The theorem below shows that this extension functor is given by an obvious universal property.

Construction of P^* (B.15)

Let $P \in {}^{\Delta}\mathcal{V}$ and $X \in \mathcal{S}$. Define a cosimplicial-simplicial object $X \square P : (\Delta^{\mathrm{op}})^{\mathrm{op}} \times \Delta^{\mathrm{op}} \to \mathcal{V}$ by $(X\square P)_n^k = \mathrm{Hom}(X_k, P_n) = P_n^{X_k}$ and obviously on morphisms. Then $P^*(X)$ is the (essentially unique) *diagonal limit* (= "end" [Mc]) of $X\square P$. I.e. we have a map $\omega : P^*(X) \to \prod_{n=0}^{\infty} (X\square P)_n^n \in \mathcal{V}$ which is universal with the property of being dinatural. Here $f : W \to \prod_{n=0}^{\infty} (X\square P)_n^n \in \mathcal{V}$ is *dinatural* if for all $\alpha : [n] \to [k] \in \Delta$ one has $(X\square P)_n^{\alpha} f_n = (X\square P)_{\alpha}^k f_k : W \to (X\square P)_n^k$. Of course ω universal means that any such f factors uniquely through ω in $\mathcal{V}$.

We obviously regard $(X,P) \mapsto \hat{P}(X) := \prod_{n=0}^{\infty} (X\square P)_n^n$ and thus $(X,P) \mapsto P^*(X)$ as a functor $\mathcal{S}^{\mathrm{op}} \times {}^{\Delta}\mathcal{V} \to \mathcal{V}$. Then P^* is a cofunctor, and a simplicial map $P \to Q$ yields a natural map $P^* \to Q^*$.

THEOREM (B.16)

We may identify $P^*(\Delta[n]) = P_n$ and analogously for the maps between the various $\Delta[k]$, i.e. $|P^*| = P$.

For any X the canonical map $\Phi : P^*(X) \to \hat{P}(X)$ is a diagonal limit of $X\square P$.

Moreover, every simplicial map $|F| \to P$ extends uniquely to a natural map $F \to P^*$.

In particular, the extension functor is right adjoint - right inverse to the restriction functor.

Proof. Recall (2.44) the definition of the natural map $\Phi : F \to \hat{F}$ for any cofunctor $F : \mathcal{S} \to \mathcal{V}$. We have

$$\hat{F}(X) = \prod_{n=0}^{\infty} F(\Delta[n])^{X_n} = \prod_{n=0}^{\infty} (X\square|F|)_n^n$$

and $\Phi_X : F(X) \to \hat{F}(X)$ is given by $\pi_\sigma \circ \Phi_X = F(\sigma_k)$ where π_σ denotes the projection onto the factor $F(\Delta[k])$ with index $\sigma \in X_k$ (also regarded as a map $\Delta[k] \to X$). One checks that Φ_X is dinatural.

The first assertion follows from the observation that the (similarly defined) map $P_n \to \hat{P}(\Delta[n])$ is a diagonal limit of $\Delta[n]\square P$, where $\hat{P}$ is defined in (B.15). In particular, we have $(P^*)\hat{} = \hat{P}$. The second assertion is another straightforward checking of the universal property. The last assertion follows. □

Remarks (B.17)

We claim that P^* is adjoint to $\mathcal{V}(-, P)$ where the simplicial set $\mathcal{V}(W, P)$ is defined in the obvious way. This means there is a natural bijection $\mathcal{V}(W, P^*(X)) \approx \mathcal{S}(X, \mathcal{V}(W, P))$ which sends f to $f' : \sigma_n \mapsto P^*(\sigma_n) \circ f$. In order to prove it observe that both sides are in natural bijection with $\text{dinat}(W, \hat{P}(X))$.

Definition (B.18)

A cofunctor $F \in {}^{\mathcal{S}}\mathcal{V}$ is *extensional* if the natural map $\eta^{(F)} : F \to |F|^*$ is isomorphic. Equivalently, F is isomorphic to some extension P^*.

COROLLARY (B.19)

The restriction functor and the extension functor constitute an equivalence

$$\langle \text{extensional cofunctors } \mathcal{S} \to \mathcal{V} \rangle \sim \langle \text{simplicial objects over } \mathcal{V} \rangle\,.$$

□

Recall that a cofunctor $F : \mathcal{S} \to \mathcal{V}$ is *representable* (with models $\Delta[n]$, as always) iff $\Phi_X : F(X) \to \hat{F}(X)$ has a natural retraction.

LEMMA (B.20)

A cofunctor $F : \mathcal{S} \to \mathcal{V}$ is extensional iff $\Phi_X : F(X) \to \hat{F}(X)$ is a diagonal limit of $X\square|F|$.

A representable cofunctor $F : \mathcal{S} \to \mathcal{V}$ is extensional.

Proof. The first assertion follows easily from the second of (B.16).

By definition the natural map $\eta : F \to |F|^*$ is unique with $\Phi'\eta = \Phi$, where $\Phi_X = \Phi_X^{(F)} : F(X) \to \hat{F}(X)$ is dinatural and $\Phi'_X = \Phi_X^{(|F|^*)} : |F|^*(X) \to \hat{F}(X)$ is universally dinatural (i.e. a diagonal limit); see above. One checks that $\hat{\eta} = 1 : \hat{F} \to \hat{F}$.

Let F be representable, so there is a natural retraction Ψ of Φ. Putting $\lambda = \Psi\Phi'$ we have $\lambda\eta = 1$ and hence $\hat{\lambda} = 1$. We conclude that $\Phi'(\eta\lambda) = (\eta\lambda)\hat{}\Phi' = \hat{\eta}\hat{\lambda}\Phi' = \Phi'$ which implies $\eta\lambda = 1$ by the universal property of Φ'. Hence η is isomorphic which means that F is extensional. □

Remark (B.21)

Up to here everything holds analogously when Δ is replaced by any small category I. The models in the category ${}^{\mathrm{I}}\langle\mathrm{Set}\rangle$ of I^{op}-diagrams are the $\mathrm{I}[j] = \mathrm{I}(-,j)$. By Yoneda the functor $\mathrm{I}[-] : \mathrm{I} \to {}^{\mathrm{I}}\langle\mathrm{Set}\rangle$ is a full embedding. Moreover $X_j = {}^{\mathrm{I}}\langle\mathrm{Set}\rangle(\mathrm{I}[j], X)$ for any $j \in \mathrm{I}$ and I^{op}-diagram X.

Recall that a simplicial object P (over $\mathcal{V}$) is *contractible* if P regarded as a chain complex is contractible, i.e. acyclic rel 0 (equivalently NP is contractible).

THEOREM (B.22)

If F is representable, then $|F|$ is contractible.

Proof. Notice that $\hat{F}$ is the product of the cofunctors $X \mapsto \mathrm{Hom}(X_n, |F|_n)$. Since a retract of a product of contractible chain complexes is contractible, the same holds for simplicial objects (over $\mathcal{V}$). Hence we may assume that F is a non-normalized chain cofunctor: $F(X) = \mathrm{Hom}(X_n, W)$. Then $|F| = \mathrm{Hom}(\Delta[-]_n \otimes \mathbb{Z}, W)$ and the assertion follows from the well-known fact that $\Delta[-]_n \otimes \mathbb{Z}$ is a contractible cochain complex, where the differential is $\delta = \sum(-1)^i \delta^i$. The desired contraction of $|F|$ is induced via $\mathrm{Hom}(-, W)$. □

LEMMA (B.23)

The normalized cochain functor $C^n(-;W)$ is representable, and $|C^n(-;W)| \cong (n, W)$ i.e. the essentially unique object of ${}^{\Delta}\mathcal{V}$ such that $N(n,W) = (\ldots 0 \leftarrow W = W \leftarrow 0 \ldots)$ with W in degree n, $n+1$.

By (B.20) it follows that $C^n(-;W) = |C^n(-;W)|^* = (n,W)^*$.

Proof. $C^n(X;W)$ is a natural retract of $\mathrm{Hom}(X_n, W) =: F(X)$, hence for the first assertion it suffices to show the latter representable with the single model $\Delta[n]$. This means that $\Phi_X : \mathrm{Hom}(X_n, W) \to \mathrm{Hom}(X_n, \mathrm{Hom}(\Delta[n]_n, W)) = \mathrm{Hom}(X_n \times \Delta[n]_n, W)$ has a natural retraction. This follows from the fact that fixing an element of $\Delta[n]_n$ yields a natural section of the projection $X_n \times \Delta[n]_n \to X_n$ (which induces Φ_X).—For the second assertion we have to compute the normalization of $\mathrm{Hom}(C_n(\Delta[-];\mathbb{Z}), W)$ which is isomorphic to $\mathrm{Hom}(-,W)$ applied to the co-normalization $N^{\vee}C_n(\Delta[-];\mathbb{Z})$ since that cofunctor carries colimits to limits. By a straightforward computation, this co-normalization is isomorphic to the cochain complex $N(n,\mathbb{Z})^* = (\ldots 0 \to \mathbb{Z} = \mathbb{Z} \to 0 \ldots)$ with $\mathbb{Z}$ in degrees n, $n+1$. It is clear that $\mathrm{Hom}(N(n,\mathbb{Z})^*, W) = N(n,W)$. □

THEOREM (B.24)

When $P \in {}^{\Delta}\mathcal{V}$ is contractible, then P^* is representable.
In fact, $P^* = \prod_{n=0}^{\infty} C^n(-;Z_n)$ where $Z_n = Z_n(NP)$.

Proof. Since P is contractible we have $NP = \prod_{n=0}^{\infty} N(n, Z_n)$. Hence P^* is the product in ${}^{\mathcal{S}}\mathcal{V}$ of the cofunctors $(n, Z_n)^*$. Since $C^n(-;Z_n) = (n,Z_n)^*$ is representable, so is P^*. □

COROLLARY (B.25)

The restriction functor and the extension functor constitute an equivalence

$$\langle\text{representable cofunctors } \mathcal{S} \to \mathcal{V}\rangle \sim \langle\text{contractible simplicial objects over } \mathcal{V}\rangle\,.$$
□

Let $\mathcal{V}$ be additive monoidal. We ask to which extend the extension of simplicial objects is compatible with tensor products.

Compatibility with tensor products (B.26)

Denote the obvious tensor product in ${}^{\mathcal{S}}\mathcal{V}$ and ${}^{\Delta}\mathcal{V}$ by $\otimes$. These are additive monoidal categories. (Identify the corresponding algebras!) The identity maps $|F \otimes G| = |F| \otimes |G|$ constitute a (commutative) monoidal transformation. We get the "adjoint" (commutative) monoidal transformation $\phi : P^* \otimes Q^* \to (P \otimes Q)^*$, $u : K \to K^*$. (These are simply the extensions of the corresponding identity maps.) From this it follows that the adjoint pair $|-|$, $-^*$ induces (adjoint) functors between the corresponding categories of algebras; cf. § 3, N° 3 and App. B, N° 1.

N° 4 Cohomology theories

Let $\mathcal{V}$ be additive monoidal. We apply (B.26) to the additive monoidal category $\langle \mathrm{dg}^* \rangle$ of cochain complexes over $\mathcal{V}$. Then we obtain functors

$$|-| : {}^{\mathcal{S}}\langle \mathrm{dg}^*\mathrm{A} \rangle \rightleftarrows {}^{\Delta}\langle \mathrm{dg}^*\mathrm{A} \rangle : -^*$$

where extension is (still) right adjoint - right inverse. In particular, the extension of $A \in {}^{\Delta}\langle \mathrm{dg}^*\mathrm{A} \rangle$ is a cofunctor $A^* : \mathcal{S} \to \langle \mathrm{dg}^*\mathrm{A} \rangle$.

Note that the degree n component $(A^*)^n$ of such an extension is the extension $(A^n)^*$ of the degree n component of A. This follows from the analogous (trivial) property of the restriction.

A^* as an adjoint cofunctor (B.27)

Given $A \in {}^{\Delta}\langle \mathrm{dg}^*\mathrm{A} \rangle$ the cofunctors

$$A^* : \mathcal{S} \leftrightarrows \langle \mathrm{dg}^*\mathrm{A} \rangle : \|-\|$$

are adjoint where $\|B\| = \mathrm{dg}^*\mathrm{A}(B, A)$. The desired adjunction

$$\mathrm{dg}^*\mathrm{A}(B, A^*(X)) \approx \mathcal{S}(X, \|B\|)$$

is a restriction of the bijection $\mathrm{dg}^*(B, A^*(X)) \approx \mathcal{S}(X, \mathrm{dg}^*(B, A))$ mentioned in (B.17). We leave the verification to the reader. In case $A_0 = K$, a base point $\Delta[0] \to X$ yields an augmentation of $A^*(X)$, and an augmentation $B \to K$ yields a basepoint of $\|B\|$ by adjunction. Hence we also get adjoint cofunctors

$$A^* : \mathcal{S}_0 \leftrightarrows \langle \mathrm{dg}^*\mathrm{A} \rangle_0 : \|-\| .$$

Definition (B.28)

An *additive cohomology theory* (with coefficient $W \in \mathcal{V}$) is a cofunctor $F : \mathcal{S} \to \langle \mathrm{dg}^* \rangle$ with unit $\eta : W \to F$ such that

(i) F is acyclic on models rel η,

(ii) F is degreewise representable.

A *cohomology theory* is a cofunctor $F : \mathcal{S} \to \langle \mathrm{dg}^*\mathrm{A} \rangle$ whose underlying cofunctor into $\langle \mathrm{dg}^* \rangle$ is an additive cohomology theory with unit $\eta : K \to F$ (the unique natural algebra map).

A simplicial object R is *simplicially trivial* if all its face and degeneracy operators are isomorphisms in $\mathcal{V}$. Then R is canonically isomorphic to the constant simplicial object R_0.

The following is an immediate consequence of our results.

THEOREM (B.29)

An additive cohomology theory gives and is given by an object $A \in {}^{\Delta}\langle \mathrm{dg}^* \rangle$ *such that*

(i) *the complex* A_n *is acyclic rel* $Z^0 A_n$, *and* $Z^0(A)$ *is simplicially trivial,*

(ii) *the simplicial object* A^k *is contractible.*

A cohomology theory gives and is given by an object $A \in {}^{\Delta}\langle \mathrm{dg}^* \mathrm{A} \rangle$ *satisfying* (i) *and* (ii). □

Remark (B.30)

In case $\mathcal{V} = \mathcal{M}od_{\mathsf{K}}$ Cartan ([Car]) defines a cohomology theory to be a simplicial object A satisfying (i) and $\pi_*(A^k) = 0$. The latter follows from (ii) and conversely if A^k is degreewise projective. Part of the theorem below is proved by Cartan using simplicial methods.

A *morphism* of additive cohomology theories is a natural cochain map $f : F \to G$ together with a map $\phi : W \to U$ between the corresponding coefficient objects commuting with the units η. A *morphism* of cohomology theories is a morphism of the underlying additive cohomology theories such that $\phi = \mathrm{id}_K$. (It is *not* assumed that f is an algebra map!)

THEOREM (B.31)

(a) *Let* F, G *be additive cohomology theories. Then any map between the coefficient objects* $\phi : W \to U$ *extends to a morphism* $f : F \to G$. *Two morphisms extending* ϕ *are naturally cochain homotopic.*

(b) *Let* F, G *be cohomology theories. Then there is a natural cochain homotopy equivalence* $f : F \rightleftarrows G : g$ *where* f, g *are unital. Moreover* f *is multiplicative up to homotopy, i.e. there is a natural cochain homotopy* $f \circ m_F \simeq m_G \circ (f \otimes f)$.

(c) *Suppose in addition that* G^n *is representable with the single model* $\Delta[n]$ *and that* $F^{n+1}(\Delta[n]) = 0$. *Then* f *exists uniquely. Moreover,* f *is multiplicative in degree* 0.

Proof. This is an immediate consequence of the method of acyclic models, i.e. the dual of (2.41) and (2.42) (resp. a slight generalization). Notice that in (a) it is actually sufficient that F satisfies (i) and G (ii). Then the last assertion of (b) follows from the fact that F acyclic on models rel η implies the same for $F \otimes F$. In fact, if $\xi : 1 \simeq \eta\rho$ is the cochain homotopy on $F(\Delta[n])$ then $\xi \otimes 1 + \eta\rho \otimes \xi : 1 \simeq \eta\rho \otimes \eta\rho$ on $F(\Delta[n]) \otimes F(\Delta[n])$. □

We do not claim that $F \otimes F$ is a cohomology theory. This cofunctor is degreewise representable, *but* with models $\Delta[n] \sqcup \Delta[m]$ $(n, m \geqslant 0)$. To see this apply (2.34) and (2.36) to the adjoint functors $\sqcup : \mathcal{T} \times \mathcal{T} \rightleftarrows \mathcal{T} : \Delta$.

Note that the assumption on G in (c) is satisfied by C^* which is the standard example of a cohomology theory.

Chapter 3

Complete Algebra

Unless otherwise specified, K is any commutative ring; and modules, tensor products, etc. are over this ring. The theory presented here extends readily to the DG context, except when (complete) groups are involved.

§ 1 Complete Augmented Algebras

Here we give most of the basic definitions used in this chapter. The main results of this section are in N° 3 where we explore the structure of the (completion of the) group algebra of a free group.

N° 1 Ring systems

We introduce the notion of a ring system, depending on the given ground ring K. When K is the ring of integers, this notion is due to Dwyer and (implicitly) to Lazard. If one is only interested in the case where the ground ring is an algebra over the rationals, then ring systems are not needed.

Let Π be a set of primes. A Π-*integer* is an element of the multiplicative subset generated by $1 \cup \Pi$ in $\mathbb{Z}$. A module X is Π-*uniquely-divisible* if multiplication with any Π-integer is bijective (thus an automorphism of X). These modules form a full subcategory $\Pi^{-1}\mathcal{M}od_{\mathsf{K}} \hookrightarrow \mathcal{M}od_{\mathsf{K}}$. Notice that if in a short exact sequence of modules $X'' \hookrightarrow X \twoheadrightarrow X'$ two modules are in $\Pi^{-1}\mathcal{M}od_{\mathsf{K}}$, so is the third. Further, $X \otimes Y$ is in $\Pi^{-1}\mathcal{M}od_{\mathsf{K}}$ if one of the factors is so.

Theorem (3.1)

There is a functor $\Pi^{-1} : \mathcal{M}od_{\mathsf{K}} \to \Pi^{-1}\mathcal{M}od_{\mathsf{K}}$ *left adjoint - left inverse to the embedding.*

This functor is exact. The canonical map $\Pi^{-1}(X \otimes Y) \to X \otimes \Pi^{-1}Y$ *is an isomorphism. In particular,* $\Pi^{-1}X = X \otimes \Pi^{-1}\mathsf{K}$.

Proof. Let $\Pi^{-1}\mathbb{Z}$ be the subring of the rationals generated by the reciprocals of the primes in Π; and let $\Pi^{-1}\mathsf{K} = \mathsf{K} \otimes \Pi^{-1}\mathbb{Z}$, which is clearly in $\Pi^{-1}\mathcal{M}od_{\mathsf{K}}$ and a ring. It suffices to show that the functor $\Pi^{-1} := - \otimes \Pi^{-1}\mathsf{K}$ is left adjoint, which amounts to the canonical identification

$$(*) \qquad \Pi^{-1}\mathcal{M}od_{\mathsf{K}} = \mathcal{M}od_{\Pi^{-1}\mathsf{K}}\,.$$

Over the ring of integers this is well-known. Hence a Π-uniquely-divisible K-module is the same as a K-module which is also a $\Pi^{-1}\mathbb{Z}$-module; but the latter is easily seen to be the same as a $\Pi^{-1}\mathsf{K}$-module. Whence the desired identification on objects.

Every element of $X \otimes \Pi^{-1}\mathsf{K} = X \otimes_{\mathbb{Z}} \Pi^{-1}\mathbb{Z}$ is of the form $\lambda x := \lambda(x \otimes 1)$, where $x \in X$ and $\lambda \in \Pi^{-1}\mathbb{Z}$. Hence any K-linear map between $\Pi^{-1}\mathsf{K}$-modules is automatically $\Pi^{-1}\mathsf{K}$-linear. □

Remarks. Similarly, the tensor product over K of two $\Pi^{-1}\mathsf{K}$-modules is canonically identified with their tensor product over $\Pi^{-1}\mathsf{K}$. Notice that (on objects) the identification $(*)$ means that a module has at most one $\Pi^{-1}\mathsf{K}$-module structure, and it has one iff it is Π-uniquely-divisible.

COROLLARY (3.2)

(a) *If X is a flat module, so is $\Pi^{-1}X$.*

(b) *Π^{-1} is a monoidal functor.*

(c) *For another set of primes $\bar{\Pi}$ we have $\bar{\Pi}^{-1}\Pi^{-1} = (\bar{\Pi} \cup \Pi)^{-1}$.* □

Note that a module X is Π-uniquely-divisible *iff* the canonical map $X \to \Pi^{-1}X$ is bijective. In general, the kernel of $X \to \Pi^{-1}X$ consists of all x such that $ax = 0$ for some Π-integer a. This is the Π-*torsion* of X. Hence X is Π-*torsion* (resp. Π-*torsion-free*) iff $X \to \Pi^{-1}X$ is zero (resp. injective).

These notions only depend on the abelian group structure of X.

Definitions (3.3)

Let $\Pi_0 \subset \Pi_1 \subset \Pi_2 \subset \dots$ be an increasing sequence of sets of primes, where $\Pi_0 = \emptyset$. The pair K, Π_* determines a sequence of ring maps

$$(*) \qquad \mathsf{K} = \mathsf{K}_0 \to \mathsf{K}_1 \to \mathsf{K}_2 \to \dots \qquad \text{where} \quad \mathsf{K}_n = \Pi_n^{-1}\mathsf{K}\,.$$

Any sequence of rings which is obtained in *this* way, is called a *ring system*.

For example, $\mathsf{K}_n = \mathsf{K}$ for all n defines the *constant* ring system.

Convention (3.4)

Different choices of a sequence Π_* may give the same ring system. We agree that (after a ring system is fixed) Π_n denotes the *minimal* set of primes such that $\mathsf{K}_n = \Pi_n^{-1}\mathsf{K}$. This plays a role in a few situations, where it is more convenient to use the sequence of primes rather than the sequence of rings.

The following definition is of great importance in § 2 and § 3. The image of an integer k under the ring map $\mathbb{Z} \to \mathsf{K}$ is denoted by $\bar{k}$.

Definition (3.5)

We call the ring system K_* *mild*, if in the ring K_n the elements $\bar{1}, \dots, \bar{n}$ are invertible. — This is equivalent to K_n being M_n-uniquely-divisible where M_n = primes $\leqslant n$.

Remarks. Every mild ring system is obtained if we choose *any* K and Π_n = *primes* $\leqslant n$. *Every constant* mild ring system is obtained if we choose *any* $\mathbb{Q}$-*algebra* K and $\Pi_n = \emptyset$.

In the latter case we actually work over a single ring, as usual. When K is a $\mathbb{Z}/p\mathbb{Z}$-algebra (a case of minor interest), the "mild" condition implies $\mathsf{K}_n = \{0\}$ for $n \geqslant p$. Note that in the ring $\{0\}$ all elements are invertible.

N° 2 Complete modules

Let $\langle \mathrm{fil} \rangle$ be the category of *filtered modules*, by which we mean *decreasing* filtered modules. Hence a filtered module V is a module (also denoted by V) together with a sequence of submodules $V = F^0V \supset F^1V \supset F^2V \supset \ldots$ The sequence of modules $\mathrm{gr}^n V = F^nV/F^{n+1}V$ $(n \geqslant 0)$ is the *associated graded module* $\mathrm{gr}\, V$.

The morphisms of $\langle \mathrm{fil} \rangle$ are given by the following definition: when V, W are filtered modules, a *filtered map* $f : V \to W$ is a module map satisfying $f(F^nV) \subset F^nW$ for all n. There is an associated graded map $\mathrm{gr}\, f$.

Special filtered maps (3.6)

Let $f : V \to W$ be a module map. Given a filtration of W there is an *induced* filtration of V as follows: $F^nV = f^{-1}(F^nW)$. Given a filtration of V (and assuming f surjective), there is a *produced* filtration of W as follows: $F^nW = f(F^nV)$. In both cases, f becomes a filtered map.

Now let f be a filtered map. We call it *inducing* if V has the induced fitration, and *producing* if W has the produced filtration. We use *normal injective* for injective and inducing; *normal surjective* for surjective and producing; and *normal bijective* for both jointly. We call f *normal* if it is a composite of a normal surjection followed by a normal injection.

The normal bijections are precisely the isomorphism in $\langle \mathrm{fil} \rangle$. These are also called *filtered isomorphisms*. For further discussion of these notions, see App. C, N° 2.

Given a ring system, we define a notion of 'complete module' which is short for 'complete filtered module'.

Definition (3.7)

Let a ring system be given. Then $W \in \langle \mathrm{fil} \rangle$ is a *complete module* if

(i) $\mathrm{gr}^m W$ is a K_m-module, for all m;

(ii) $W = \lim\{W/F^kW\}$.

The complete modules form a full subcategory $\langle \mathrm{cpl} \rangle$ of $\langle \mathrm{fil} \rangle$.

Suppose that for $V \in \langle \mathrm{fil} \rangle$ there is a filtered map $c : V \to \widehat{V}$ which is universal among all filtered maps into a complete module. Then we call any such $\widehat{V}$ a *completion* of V, and c the *completion map*. (We show below that every filtered module has a completion.)

Lemma (3.8)

Let W be a filtered module satisfying (ii). *Then* (i) *is equivalent to*

(i′) *F^mW is a K_m-module, for all m.*

Proof. The implication (i′)⇒(i) is trivial and makes no use of (ii). For the converse note that (ii) implies $F^mW = \lim_{k \geqslant m}\{F^mW/F^kW\}$. □

Theorem (3.9)

A map f in $\langle \mathrm{cpl} \rangle$ is normal bijective (resp. *normal injective, normal surjective*) *iff* $\mathrm{gr}\, f$ *is bijective* (resp. *injective, surjective*). □

For a proof, see App. C, N° 2. The only property of complete modules needed here is (ii).

Topological considerations (3.10)

It is well-known that a filtration of a group induces a topology compatible with the group structure. We do not use this here, but give *ad hoc* definitions of some topological notions, and some elementary facts. We concentrate on the abelian case, but all one needs is a group filtered by normal subgroups.

We call a filtered module W *separated* if in the exact sequence $\bigcap_k F^k W \to W \twoheadrightarrow \lim\{W/F^k W\}$ the first module is zero. We call it *Hausdorff-complete* if (in addition) the last map is bijective.

Given elements $x_n \in W$ $(n \geqslant 0)$, we regard $\sum_{n=0}^{\infty} x_n$ as the sequence (in W) of its finite partial sums. We call such a sequence *generic* if $x_n \in F^n W$. A *limit* of a generic sequence is an element $x \in W$ such that $x \equiv \sum_{n=0}^{k} x_n \mod F^{k+1}W$ for all k. In case a unique limit exists, we use $\sum_{n=0}^{\infty} x_n$ to denote both a generic sequence and its limit.

A filtered map $f : V \to W$ is *continuous* in the sense that if x is a limit of a generic sequence $\sum x_n$, then $f(x)$ is a limit of $\sum f(x_n)$. We leave the following as a straightforward exercise to the reader:

A filtered module W is separated (resp. Hausdorff-complete) iff any generic sequence has at most (resp. precisely) one limit.

THEOREM (3.11)

Let X be a submodule of the complete module W, denote by $\widehat{X}$ the submodule consisting of the limits (in W) of all generic sequences $\sum \lambda_n x_n$ with $\lambda_n \in \mathsf{K}_n$, $x_n \in X \cap F^n W$.

(a) *If both are endowed with the induced filtration, $\widehat{X}$ is a completion of X.*

(b) *If X is a complete submodule, i.e. $X = \widehat{X}$, then W/X with the produced filtration is complete.*

Proof. *Ad* (a). We have to check the universal property. Let $f : X \to U$ be a filtered map into a complete module. To $\sum_{n=0}^{\infty} \lambda_n x_n \in \widehat{X}$ we relate $\sum_{n=0}^{\infty} \lambda_n f(x_n) \in U$. If that element of $\widehat{X}$ may also be written as $\sum_{n=0}^{\infty} \lambda'_n x'_n$, then $\sum_{n=0}^{m} \lambda_n x_n - \lambda'_n x'_n$ lies in $F^{m+1}W$, hence in $F^{m+1}X \otimes \mathsf{K}_m$ (which we may regard as a submodule of the K_m-module $F^{m+1}W$ since K_m is flat). Since f is a filtered map, we may conclude that $\sum_{n=0}^{m} \lambda_n f(x_n) - \lambda'_n f(x'_n)$ lies in $F^{m+1}U$; hence $\sum_{n=0}^{\infty} \lambda_n f(x_n) = \sum_{n=0}^{\infty} \lambda'_n f(x_n)'$ in U, and the above relation is a well-defined map $\widehat{f} : \widehat{X} \to U$. It is clearly a filtered map extending f. Given another such map g, we have $g(\sum_{n=0}^{\infty} \lambda_n x_n) \equiv g(\sum_{n=0}^{m} \lambda_n x_n) = \sum_{n=0}^{m} \lambda_n f(x_n) \mod F^{m+1}W$, hence $g = \widehat{f}$.

Ad (b). By definition of the filtrations, $X \hookrightarrow W \twoheadrightarrow W/X$ is short exact in each filtration (cf. App. C, N° 2). Hence $F^n(W/X)$ is a K_n-module. It is clear that every generic sequence in W/X has a limit, so it remains to prove it separated. If y lies in each filtration of W/X, then we find an inverse image $w_n \in F^n W$ for all n. Hence $x_n = w_n - w_{n+1} \in F^n X$ and $w_0 = \sum x_n \in X$ because X is complete. But $w_0 \mapsto y$, so we may conclude that $y = 0$. □

Remarks (3.12)

(a) If X has the property that $F^m X = X \cap F^m W$ is a K_m-module, then $\widehat{X}$ is the submodule consisting of the limits of all generic sequences in X. (This is called the *closure* of X, and X is said to be *closed* if it coincides with its closure). Hence X is a complete submodule iff $F^m X$ is a K_m-module and X is closed.

(b) As an example consider any filtered module V and let $c : V \to \widehat{V}$ be a completion map. It is easy to see that $\mathrm{Im}(c) \hookrightarrow \widehat{V}$ (induced filtration) is also a completion. Hence any element of $\widehat{V}$ is the limit of a generic series $\sum \lambda_n x'_n$ where $\lambda_n \in \mathsf{K}_n$, $x'_n \in \mathrm{Im}(c)$. One may show that $V \to \mathrm{Im}(c)$ is a filtered isomorphism, if V is separated and $\mathrm{gr}^n V$ Π_n-torsion-free, cf. (C.11) (b).

THEOREM (3.13)

There is a functor $\widehat{\ }$: $\langle \mathtt{fil} \rangle \to \langle \mathrm{cpl} \rangle$ *left adjoint - left inverse to the embedding.* In other words: *Any filtered module has a completion.*

Proof. We define $\widehat{V}$ as a quotient module of $P = \prod_{n=0}^{\infty} F^n V \otimes \mathsf{K}_n$, which we regard as the set of all formal sums $\sum_{n=0}^{\infty} y_n$ where $y_n \in F^n V \otimes \mathsf{K}_n$. (We also denote by y_n its image in $V \otimes \mathsf{K}_m$, $m \geqslant n$). We factor out the submodule of all formal sums with the property:

(N) for all $m \geqslant 0$, the element $\sum_{n=0}^{m} y_n \in V \otimes \mathsf{K}_m$ lies in $F^{m+1} V \otimes \mathsf{K}_m$.

Let $F^k \widehat{V}$ be the submodule given by all elements of the form $\sum_{n=k}^{\infty} y_n$. Then the obvious map $c : V \to \widehat{V}$ preserves filtration, and $\widehat{V}$ is complete. The latter follows from (3.11). In fact, the canonical (complete) filtration of P produces the filtration of $\widehat{V}$, and the kernel of $P \to \widehat{V}$ is a complete submodule. The proof of the universal property of c is essentially the same as in (3.11) (b). □

In the following we use notation from (3.19). We show that a filtered map (into a Hausdorff-complete module) is a completion iff the associated graded map is a completion.

PROPOSITION (3.14)

A filtered map $g : V \to W$ *is a completion of* V *iff it has the property*

(∗) *the target of* g *is complete, and* $\mathrm{gr}(g) \otimes \mathsf{K}_*$ *is an isomorphism.*

Proof. Notice that the completion of V is unique up to canonical isomorphism. Hence, as c satisfies (∗) by construction, so does *any* completion map. Conversely, if g satisfies (∗), the universal property of c yields a canonical filtered map $\widehat{g} : \widehat{V} \to W$ in $\langle \mathrm{cpl} \rangle$. Since also c satisfies (∗), $\mathrm{gr}(\widehat{g})$ is an isomorphism. The assertion follows thus from (3.9). □

Example (3.15)

If V is *splittable*, i.e. there are module splittings $V \cong F^{n+1} V \oplus V_n$ for all n, then $\prod_{n \geqslant 0} \mathrm{gr}^n V \otimes \mathsf{K}_n$, with the *canonical* filtration $F^k = \prod_{n \geqslant k} \mathrm{gr}^n V \otimes \mathsf{K}_n$, is a completion of V. (The completion map is canonical in terms of the given splittings). In particular, $\widehat{V}$ is again splittable.

Next we discuss the 'tensor product' of filtered modules; in a standard way it yields the so-called 'complete tensor product' of complete modules.

Definitions (3.16)

The *tensor product* of filtered modules V, W is the module $V \otimes W$ with the filtration $F^k(V \otimes W) = \mathrm{Im}(\bigoplus_{i+j=k} F^i V \otimes F^j W \to V \otimes W)$.

The *complete tensor product* of complete modules V, W is then the complete module $V \mathbin{\widehat{\otimes}} W = \widehat{V \otimes W}$. We thus have the completion map $c : V \otimes W \to V \mathbin{\widehat{\otimes}} W$, and write $c(x \otimes y) =: x \mathbin{\widehat{\otimes}} y$.

In this way $\langle \mathrm{fil} \rangle$, resp. $\langle \mathrm{cpl} \rangle$, become strongly additive (B.3) monoidal categories.

PROPOSITION (3.17)

Let V, W be filtered modules. The canonical graded map $\pi : \mathrm{gr}(V) \otimes \mathrm{gr}(W) \to \mathrm{gr}(V \otimes W)$ is surjective. □

This is easily seen. The map π is *not* necessarily bijective, but it *is* so on certain conditions, see (C.24).

THEOREM (3.18)

Let V, W be filtered modules. The canonical filtered map $V \otimes W \to \widehat{V} \mathbin{\widehat{\otimes}} \widehat{W}$ is a completion.

Hence the completion functor is monoidal.

Proof. Let $V_n = V/F^{n+1}V$, endowed with the filtration produced by $q : V \to V_n$. In (C.19) we show that the filtered map $q \otimes q : V \otimes W \to V_n \otimes W_n$ has the property that $\mathrm{gr}(q \otimes q)$ is an isomorphism in degrees $\leqslant n$. We also apply this to the completed filtered modules. Hence it suffices to show that

$$\mathrm{gr}(\bar{c} \otimes \bar{c}) \otimes \mathsf{K}_n : \mathrm{gr}(V_n \otimes W_n) \otimes \mathsf{K}_n \longrightarrow \mathrm{gr}(\widehat{V_n} \otimes \widehat{W_n}) \otimes \mathsf{K}_n$$

is an isomorphism, where $\bar{c}$ is the filtered map induced by c. Since K_n is flat, $\mathrm{gr}(\bar{c} \otimes \bar{c}) \otimes \mathsf{K}_n = \mathrm{gr}(\bar{c} \otimes \bar{c} \otimes \mathsf{K}_n)$; and since $\mathsf{K}_n \otimes \mathsf{K}_n = \mathsf{K}_n$, we have $\bar{c} \otimes \bar{c} \otimes \mathsf{K}_n = \bar{c} \otimes \mathsf{K}_n \otimes \bar{c} \otimes \mathsf{K}_n$. Hence we are reduced to showing that

$$\bar{c} \otimes \mathsf{K}_n : V_n \otimes \mathsf{K}_n \longrightarrow \widehat{V_n} \otimes \mathsf{K}_n$$

is an isomorphism. But these filtered modules are discrete, thus Hausdorff-complete, and the associated graded of this map is the isomorphism $\mathrm{gr}^{\leqslant n}(V) \otimes \mathsf{K}_n = \mathrm{gr}^{\leqslant n}(\widehat{V}) \otimes \mathsf{K}_n$. By (3.9) this completes the proof. □

Finally we indicate analogous considerations for graded modules.

Definition (3.19)

A *graded module* is a sequence of (K-)modules, indexed by (superscript) non-negative integers. (Occasionally we consider a graded module as a module together with a direct sum decomposition.) The tensor product (monoidal structure) of graded modules is defined as usual, note that here the interchange map involves no sign. A *complete* graded module has the property that its n-th component is a K_n-module. There is an obvious completion functor sending the graded module $M = \{M^n\}$ to the complete graded module

$$\widehat{M} = M \otimes \mathsf{K}_* = \{M^n \otimes \mathsf{K}_n\}.$$

Given complete graded modules M, N we clearly define their *complete tensor product* by $M \mathbin{\widehat{\otimes}} N = \widehat{M \otimes N}$.

If V is a filtered module, we have a canonical isomorphism $\mathrm{gr}(\widehat{V}) = \widehat{\mathrm{gr}}(V)$, see (3.14). Hence if V, W are complete modules, we have a canonical surjective map

$\mathrm{gr}(V)\mathbin{\widehat{\otimes}}\mathrm{gr}(W)\to\mathrm{gr}(V\mathbin{\widehat{\otimes}}W)$, cf. (3.17). It is bijective on certain conditions (e.g. if $\mathrm{gr}\,V$ is flat), cf. (C.24).

Graduations vs. Filtrations (3.20)

Given a (complete) graded module M we may construct a (complete) filtered module $\bigoplus_n M^n$ (resp. $\prod_n M^n$), where the filtration is obvious. This gives rise to a functor, T say, which is left inverse to 'gr' up to canonical isomorphism and compatible with (complete) tensor product and completion. Note that T is essentially a forgetful functor. There is a category equivalence by which a (complete) graded module can be regarded as a (complete) filtered module endowed with explicit splittings for the filtration submodules, cf. (3.15).

N° 3 Complete augmented algebras and free groups

Let $\langle \mathrm{AA}\rangle$ be the category of *augmented algebras* (unital, associative K-algebras B endowed with an algebra map $\varepsilon : B\to\mathsf{K}$). The *augmentation ideal* is $\overline{B}=\mathrm{Ker}(\varepsilon)$.

Definitions (3.21)

A *filtered augmented algebra* is an augmented algebra B which is also a (decreasing) filtered module, such that (i) $F^iB\cdot F^jB\subset F^{i+j}B$ (i.e. mult : $B\otimes B\to B$ is filtration preserving), and (ii) $F^1B=\overline{B}$.

A *complete augmented algebra* is a filtered augmented algebra which is complete as filtered module.—With the evident notion of morphism these form a category $\langle \mathrm{CAA}\rangle$.

Completion (3.22)

Let B be a filtered AA. Then the completion (see N° 2) $\widehat{B}$ is a CAA, and $c : B\to\widehat{B}$ is a filtered AA-map having the obvious universal property. The multiplication of $\widehat{B}$ is the obvious composite $\widehat{B}\otimes\widehat{B}\to\widehat{B}\mathbin{\widehat{\otimes}}\widehat{B}=\widehat{B\otimes B}\to\widehat{B}$ where we use (3.18). Since this composite is a filtered map, $\widehat{B}$ satisfies (i); and condition (ii) is easily verified.

Tensor product (3.23)

For general reasons we have a tensor product for filtered (resp. complete) AAs—one only has to check that (ii) is satisfied by the filtered tensor product of two filtered AAs. In other words, when B, B' are filtered AAs, so is $B\otimes B'$ (the filtered tensor product, the standard algebra structure). Hence, when A, A' are CAAs, so is $A\mathbin{\widehat{\otimes}}A'$.

Example (3.24)

Any augmented algebra B admits the *augmental filtration* $F^kB=\overline{B}^{\,k}$. It is minimal among all filtrations satisfying (i)–(ii) in (3.21).—One checks: If B, B' are endowed with the augmental filtration, then the tensor product filtration on $B\otimes B'$ is again the augmental filtration.

Note: When B is a filtered AA, then $\mathrm{gr}(B)$ is a graded algebra. The multiplication is the composite of the canonical map $\mathrm{gr}(B)\otimes\mathrm{gr}(B)\to\mathrm{gr}(B\otimes B)$ followed by the associated graded map $\mathrm{gr}(B\otimes B)\to\mathrm{gr}(B)$ of the multiplication of B.

Complete tensor algebra (over a module) (3.25)

For a module W consider the tensor algebra $\mathsf{T}W = \bigoplus_{n=0}^{\infty} W^{\otimes n}$ endowed with the augmental filtration, so that $F^k\mathsf{T}W = \bigoplus_{n=k}^{\infty} W^{\otimes n}$.—Its completion is the *complete tensor algebra* $\widehat{\mathsf{T}}W = \prod_{n=0}^{\infty} W^{\otimes n} \otimes \mathsf{K}_n$ with filtration $F^k\widehat{\mathsf{T}}W = \prod_{n=k}^{\infty} W^{\otimes n} \otimes \mathsf{K}_n$.—These constructions have the obvious universal property. For instance, $\widehat{\mathsf{T}}$ is left adjoint to the functor which sends any CAA to its augmentation ideal.

When W is free with basis $\mathbf{X}_S = \{\mathbf{X}_s\}_{s\in S}$, we also write $\mathsf{T}W = \mathsf{K}\langle \mathbf{X}_S\rangle$; this is the algebra of polynomials in the non-commuting indeterminates $\mathbf{X}_s$. For the completion we also write $\widehat{\mathsf{T}}W = \mathsf{K}\langle\langle \mathbf{X}_S\rangle\rangle$; this is the algebra of *formal power series* in the non-commuting indeterminates $\mathbf{X}_s$. It is also called the *Magnus algebra* on the set $\mathbf{X}_S$.

Complete group algebra of a free group (3.26)

Let F be the free group generated by the set S; the group algebra $\mathsf{K}F$ has a natural augmentation induced by $F \to \{1\}$. Except for the trivial case, $\mathsf{K}F$ is *not* free as AA. We endow it with the augmental filtration. We prove below that the completion $\widehat{\mathsf{K}}F$ is naturally CAA-isomorphic to the free CAA $\mathsf{K}\langle\langle \mathbf{X}_S\rangle\rangle$. Observe that the group algebra functor $\mathsf{K}-$ (and thus also $\widehat{\mathsf{K}}-$) has a canonical right adjoint, so it preserves coproducts.

In the following theorem we explore the structure of the group algebra of a free group.

THEOREM (3.27)

Given a set S, let F be the free group generated by S, and V the free module generated by $(\mathbf{X}_s)_{s\in S}$. We consider the AA-map

$$\theta : \mathsf{T}V \to \mathsf{K}F\,, \qquad \mathbf{X}_s \mapsto s - 1\,.$$

It has componenents $\theta^k : V^{\otimes k} \to \overline{\mathsf{K}F}^{\,k}$.

(a) *The map θ^k induces an (F-module) isomorphism*

$$\Theta^k :\ \mathsf{K}F \otimes V^{\otimes k} \xrightarrow{\ \simeq\ } \overline{\mathsf{K}F}^{\,k}\,, \qquad x \otimes u \mapsto x \cdot \theta^k(u)\,.$$

(b) *The sum of the maps $\overline{\mathsf{K}F}^{\,k+1} \subset \overline{\mathsf{K}F}^{\,k}$ and θ^k is a module isomorphism*

$$\overline{\mathsf{K}F}^{\,k+1} \oplus V^{\otimes k} \xrightarrow{\ \simeq\ } \overline{\mathsf{K}F}^{\,k}\,.$$

(c) *Considering the augmental filtrations we have*

$$\operatorname{gr}\theta :\ \mathsf{T}V \xrightarrow{\ \simeq\ } \operatorname{gr}\mathsf{K}F\,.$$

Using the formula $\operatorname{gr}\widehat{\theta} = \operatorname{gr}(\theta) \otimes \mathsf{K}_*$, the latter implies the following.

COROLLARY (3.28)

There is a natural CAA-isomorphism $\widehat{\theta} : \widehat{\mathsf{T}}V \xrightarrow{\ \simeq\ } \widehat{\mathsf{K}}F$. □

Proof. Ad (a). For a proof that $\Theta^1 : \mathsf{K}F \otimes V \to \overline{\mathsf{K}F}$ is an isomorphism, see [HiSt | p. 196]. It follows at once that $\overline{\mathsf{K}F}^{\,k+1}$ is generated as a module by $\{v(s -$

$1) \mid v \in \overline{\mathsf{K}F}^k,\ s \in S\}$. Hence the restriction of Θ^1 to $\overline{\mathsf{K}F}^k \otimes V \to \overline{\mathsf{K}F}^{k+1}$, $k \geqslant 1$, is also an isomorphism. Since $\Theta^{k+1} = \Theta^1(\Theta^k \otimes V)$ the assertion follows by induction.

One checks that under Θ^{k+1}, Θ^k the inclusion $\overline{\mathsf{K}F}^{k+1} \hookrightarrow \overline{\mathsf{K}F}^k$ corresponds to $\Theta^1 \otimes 1 : \mathsf{K}F \otimes V^{\otimes k+1} \to \mathsf{K}F \otimes V^{\otimes k}$ where $1 = 1_{V^{\otimes k}}$. By (a), $\mathrm{Coker}(\Theta^1 \otimes 1) = \varepsilon \otimes 1 : \mathsf{K}F \otimes V^{\otimes k} \to V^{\otimes k}$, thus $V^{\otimes k} \cong \mathrm{gr}^k(\mathsf{K}F)$ via θ^k proving (c). Since $\varepsilon \otimes 1$ has a canonical section we see that (b) holds. □

Power series in CAAs (3.29)

Let A be a CAA. By the universal property of the complete tensor algebra, there is for any $x \in \overline{A}$ a unique CAA-map $\mathsf{K}\langle\langle \mathbf{X} \rangle\rangle \to A$, $\mathbf{X} \mapsto x$. The image under this map of any formal power series $\mathbf{f}(\mathbf{X})$ is denoted by $\mathbf{f}(x)$. Let $\mathbf{f}(\mathbf{X}) = \sum c_n \mathbf{X}^n$, then $\mathbf{f}(x) \in A$ is the limit of the generic series $\sum c_n x^n$ in A, cf. (3.10). Using this concept one easily sees that $1 + \overline{A}$ is a group under multiplication. In fact, the formal power series $1 + \mathbf{X}$ is invertible in $\mathsf{K}\langle\langle \mathbf{X} \rangle\rangle$.

Remark (3.30)

The inverse of the isomorphism in (3.28) is induced by the monoid map $F \to \mathsf{K}\langle\langle \mathbf{X}_s \rangle\rangle_{s \in S}$, $s \mapsto 1 + \mathbf{X}_s$ which exists by (3.29). When the ring system is mild (e.g. K is a $\mathbb{Q}$-algebra), there is another isomorphism (3.45) which is even compatible with the canonical cocommutative diagonals where, by definition, s is grouplike and $\mathbf{X}_s$ primitive.

Complete tensor algebra (general case) (3.31)

Let V be a filtered module with $F^1V = V$. The *filtered tensor algebra* $\mathsf{T}V$ is the filtered direct sum of the filtered tensor products $V^{\otimes n}$ $(n \geqslant 0)$, cf. (C.7). It is clear that this is a filtered AA. — The completion $\widehat{\mathsf{T}}V$ is a CAA, the *complete tensor algebra* over V. The natural filtered map $V \to \widehat{\mathsf{T}}V$ has the obvious universal property. We have $\widehat{\mathsf{T}}V = \prod_{n=0}^{\infty} V^{\widehat{\otimes} n}$, since $V^{\otimes n}$ and so its completion coincides with its n-th filtration, cf. (C.7). The construction in (3.25) is the special case where V has the trivial filtration $(F^2 = 0)$.

In case V is a free filtered module in the sense that $V = \bigoplus_{n \geqslant 1} V^n$ with V^n a free module and $F^kV = \bigoplus_{n \geqslant k} V^n$, any CAA isomorphic to $\widehat{\mathsf{T}}V$ is said to be *free*.

N° 4 Rigidity

The condition of rigidity introduced next is assumed from the beginning in the work of Quillen and Dwyer. Although this assumption leads to some simplifications (e.g. in the simplicial context), it is not really necessary. Moreover there are applications (e.g. in the dg context) where non-rigid objects appear naturally.

We use here the complete tensor algebra over graded modules, which is the usual tensor algebra followed by completion of graded modules, see (3.19). For a more detailed discussion see §2, N° 2.

Definition (3.32)

A filtered AA B is *rigid* (over K_*) if the graded AA $\mathrm{gr}(B/F^{n+1}B) \otimes \mathsf{K}_n = \mathrm{gr}^{\leqslant n}(B) \otimes \mathsf{K}_n$ is generated by the elements of degree 1.

Examples (3.33)

The augmental filtration of any AA B has the stronger property

(∗) $\mathrm{gr}\, B$ is generated by $\mathrm{gr}^1 B$.

It follows from (3.36) that, for any filtered augmented algebra B, this condition is equivalent to $F^nB = \overline{B}^n + F^{n+1}B$ (all n); or to F^nB being the closure of $\overline{B}^n$ in B, cf. (3.12) (a). Hence when the filtration is finite (discrete case), ($*$) implies $F^nB = \overline{B}^n$. Notice that over the constant ring system rigid means ($*$).

It is clear that B is rigid iff $\widehat{B}$ is rigid. Hence the complete tensor algebra over a module (3.25) is rigid. On the other hand, one easily constructs non-rigid (free) CAAs using (3.31).

LEMMA (3.34)

Let $f : B' \to B$ be a filtered AA-map.

(a) *If B' is rigid and* $\mathrm{gr} f$ *surjective, then B is rigid;*

(b) *If B is rigid and* $\mathrm{gr}^1(f)$ *surjective, then* $\mathrm{gr} f$ *is surjective.* □

Since a surjective map f has the property that $\mathrm{gr}^1(f)$ is surjective, we deduce ...

PROPOSITION (3.35)

Let $f : A' \to A \in \langle \mathrm{CAA} \rangle$ and assume A rigid. Then f surjective implies it is normal surjective. Hence: *f is a filtered isomorphism iff it is bijective.* □

THEOREM (3.36)

The following are equivalent conditions for $A \in \langle \mathrm{CAA} \rangle$ to be rigid:

(i) *the canonical graded AA-map* $\widehat{\mathsf{T}}\mathrm{gr}^1(A) \to \mathrm{gr}(A)$ *is surjective;*

(ii) *the canonical CAA-map* $\widehat{\mathsf{T}}(\overline{A}) \to A$ *is normal surjective;*

(iii) $F^nA/(\overline{A}^n + F^{n+1}A)$ *is Π_n-torsion (all n);*

(iv) $F^nA = (\overline{A}^n)^\wedge$ *in A (all n).*

Proof. In (ii) we regard $\overline{A}$ just as a module; see (3.25). Notice that $\mathrm{gr}\widehat{\mathsf{T}}(\overline{A}) = \widehat{\mathsf{T}}(\overline{A})$ sends a natural surjection to $\widehat{\mathsf{T}}\mathrm{gr}^1(A)$, so (i)⇔(ii) is clear by (3.9). One verifies that (ii) is equivalent to

(ii′) F^nA is the set of all limits $\sum_{m=n}^{\infty} \lambda_m x_m$ where $x_m \in \overline{A}^m$ and $\lambda_m \in \mathsf{K}_m$.

The proof of (ii′)⇒(iv)⇒(iii)⇒(ii′) presents no difficulty. □

For example, let A be a rigid CAA which is Π_m-torsion-free for all m (this is always the case if the ring system is constant). Then in A the filtration is determined by the topology; further a continuous AA-map to another CAA preserves filtration.

The following result is useful if one wants to apply the machine of homotopical algebra to *simplicial* CAAs. We are not doing so, but need the completely analogous result on CLAs (§ 2, N° 4).

Notice that a rigid CAA is free, in the sense of (3.31), iff it is isomorphic to a Magnus algebra (3.25), or equivalently, to a coproduct of copies of $\mathsf{K}\langle\langle \mathbf{X} \rangle\rangle$.

THEOREM (3.37)

The category $\langle \mathrm{CAA} \rangle'$ of rigid CAAs has arbitrary limits and colimits. Moreover, the effective epimorphisms are precisely the surjective maps, and $\mathsf{K}\langle\langle \mathbf{X} \rangle\rangle$ is a projective generator.

Proof. There is an analogous result for $\langle \mathrm{CAA} \rangle$ itself, where the effective epimorphisms are the normal surjective maps, and there is a countable set $\{T_n\}_{n \geqslant 1}$ of projective generators. Namely, T_n is the free CAA on one generator in filtration n (but not $n+1$), so that $T_1 = \mathsf{K}\langle\langle \mathbf{X} \rangle\rangle$. We leave the details to the reader. — The desired result on $\langle \mathrm{CAA} \rangle'$ follows easily since the category embedding into $\langle \mathrm{CAA} \rangle$ has a right adjoint. To construct it, let $A \in \langle \mathrm{CAA} \rangle$. Define A_{rig} to be the same AA, but with filtration produced by the map in (3.36) (ii). Using (C.15) it is easy to see that A_{rig} is a rigid CAA, and that $A_{\mathrm{rig}} \to A$, $x \mapsto x$ is a co-universal CAA-map.

□

§ 2 COMPLETE LIE ALGEBRAS AND COMPLETE HOPF ALGEBRAS

For the results of this section (and the following) we need to know that, roughly speaking, in any kind of algebra "n-fold products are divisible by n". Hence we work over a *mild* ring system, see § 1, N° 1. As a special case one may choose K to be a $\mathbb{Q}$-algebra.

N° 1 Complete Hopf algebras and the exponential mapping

Fix a mild ring system K_*.

Definitions (3.38)

A *complete Hopf algebra* is an $A \in \langle \mathrm{CAA} \rangle$ endowed with a CAA-map $\Delta : A \to A \mathbin{\widehat{\otimes}} A$ (the *diagonal*) which is cocommutative, coassociative and has the unique CAA-map $A \to \mathsf{K}$ as a counit. — With the evident notion of morphism these form a category $\langle \mathrm{CHA} \rangle$.

In other words, $\langle \mathrm{CHA} \rangle$ is the category of cocommutative comonoids over the monoidal category $\langle \mathrm{CAA} \rangle$.

We introduce the adjoint functors in the following diagram where $\langle \mathrm{Gp} \rangle$ is the category of groups and $\langle \mathrm{LA} \rangle$ that of Lie algebras (over K).

$$\langle \mathrm{Gp} \rangle \underset{\mathcal{G}}{\overset{\widehat{\mathsf{K}}}{\rightleftarrows}} \langle \mathrm{CHA} \rangle \underset{\mathcal{P}}{\overset{\widehat{U}}{\leftrightarrows}} \langle \mathrm{LA} \rangle\,. \tag{3.39}$$

Let B be a (cocommutative, as always) Hopf algebra, endowed with its augmental filtration. Then the diagonal $\Delta : B \to B \otimes B$ preserves filtration (i.e. B is a filtered Hopf algebra). Hence, there is an induced diagonal $\widehat{\Delta} : \widehat{B} \to \widehat{B \otimes B} \to \widehat{B} \mathbin{\widehat{\otimes}} \widehat{B}$, and $\widehat{B}$ is a CHA.

The functors $\widehat{\mathsf{K}}$, $\widehat{U}$ (3.40)

For $G \in \langle \mathrm{Gp} \rangle$ the group algebra $\mathsf{K}G$ (resp. for $L \in \langle \mathrm{LA} \rangle$ the universal enveloping UL) is a Hopf algebra, whose diagonal is induced by $(1,1) : G \to G \times G$ (resp. $L \to L \times L$). The functors are thus defined.

The functors $\mathcal{G}, \mathcal{P}$ (3.41)

For $A \in \langle \text{CHA} \rangle$ let

$$\begin{aligned} \mathcal{G}A &= \{x \in 1 + \overline{A} \mid \Delta x = x \mathbin{\widehat{\otimes}} x\} \\ \mathcal{P}A &= \{x \in \overline{A} \mid \Delta x = x \mathbin{\widehat{\otimes}} 1 + 1 \mathbin{\widehat{\otimes}} x\}. \end{aligned}$$

Here $\mathcal{G}A$ (the set of *grouplike* elements) is a subgroup of the multiplicative group $1 + \overline{A}$, and $\mathcal{P}A$ (the set of *primitive* elements) is a sub-LA of $\overline{A}$.

It is easy to see that these functors are right adjoint to $\widehat{\mathsf{K}}$, $\widehat{U}$ respectively.

Since we work over a *mild* ring system, we have the exponential series $\exp(\mathbf{X}) \in \mathsf{K}\langle\langle \mathbf{X} \rangle\rangle$.

For any $A \in \langle \text{CAA} \rangle$, it induces a natural bijection

$$\log : 1 + \overline{A} \approx \overline{A} : \exp$$

since the inverse is given by the formal power series $\log(1 + \mathbf{X})$; cf. (3.29).

We have the formula (for $x, y \in \overline{A}$)

$$\exp(x + y) = \exp(x) \cdot \exp(y) \qquad \text{if } [x, y] = 0\,. \tag{3.42}$$

This is shown by computation; note that $z = z' \Leftrightarrow z \equiv z' \bmod F^n A$ (all n), since A is separated.

Now let $A \in \langle \text{CHA} \rangle$. Then the restrictions

$$\log : \mathcal{G}A \approx \mathcal{P}A : \exp \tag{3.43}$$

are defined. In fact:

$$\begin{aligned} x \in \mathcal{P}A \quad &\Leftrightarrow \quad \Delta(x) = x \mathbin{\widehat{\otimes}} 1 + 1 \mathbin{\widehat{\otimes}} x \\ &\Leftrightarrow \quad \exp(\Delta x) = \exp(x \mathbin{\widehat{\otimes}} 1 + 1 \mathbin{\widehat{\otimes}} x) = \exp(x \mathbin{\widehat{\otimes}} 1) \exp(1 \mathbin{\widehat{\otimes}} x) \\ &\qquad\qquad = (\exp(x) \mathbin{\widehat{\otimes}} 1)(1 \mathbin{\widehat{\otimes}} \exp(x)) = \exp(x) \mathbin{\widehat{\otimes}} \exp(x) \\ &\Leftrightarrow \quad \exp(x) \in \mathcal{G}A\,; \quad \text{here we use (3.42).} \end{aligned}$$

Definition (3.44)

We regard $\mathsf{K}\langle\langle \mathbf{X}_S \rangle\rangle$ as a CHA by defining each $\mathbf{X}_s$ to be primitive.

A rigid CHA A is *free* if $A \cong \mathsf{K}\langle\langle \mathbf{X}_S \rangle\rangle$ for some set S, or equivalently, if A is a coproduct of some copies of $\mathsf{K}\langle\langle \mathbf{X} \rangle\rangle$.

Here A is *rigid* if it is so-called as CAA. The definition of free CHA in general is similar but not needed, cf. (3.31).

Let $\mathsf{F}(S)$ be the free group (resp. $\mathsf{L}(S)$ the free LA) generated by the set S. There are canonical CHA-isomorphisms

$$\widehat{\mathsf{K}}\,\mathsf{F}(S) \xrightarrow[\varphi]{\simeq} \mathsf{K}\langle\langle \mathbf{X}_S \rangle\rangle \xleftarrow[\psi]{\simeq} \widehat{U}\,\mathsf{L}(S) \tag{3.45}$$

given by $\varphi : s \mapsto \exp(\mathbf{X}_s)$, $\psi : s \mapsto \mathbf{X}_s$.

As for ψ this is clear by $\widehat{U}\mathsf{L} = \widehat{\mathsf{T}}$ and (3.25). As for φ note that $\text{gr}(\varphi) = \text{gr}(\widehat{\theta^{-1}})$ where $\widehat{\theta^{-1}} : s \mapsto 1 + \mathbf{X}_s$, see (3.30).

Finally we mention the canonical bijection

$$\text{(3.46)} \qquad \mathrm{CHA}(\mathsf{K}\langle\langle \mathbf{X}\rangle\rangle, A) \approx \mathcal{P}A$$

defined by $f \mapsto f(\mathbf{X})$. In fact, $\mathrm{CAA}(\mathsf{K}\langle\langle \mathbf{X}\rangle\rangle, A) \approx \overline{A}$ by (3.25); and the assertion follows since $\mathbf{X}$ is primitive. One can also use (3.45) and the adjointness of $(\widehat{U}, \mathcal{P})$.

N° 2 The PBW–Theorem

We use the term 'Lie algebra' (abbreviated by 'LA') in the classical sense, except that the relation $[x,x] = 0$ is replaced by the weaker condition of anti-commutativity: $[x,y] = -[y,x]$. The latter has the advantage of being definable in categorical terms, in particular in the DG setting. Anyway, the two conditions lead to the same notion of complete Lie algebra (see below), since we work over a *mild* ring system.

Definitions (3.47)

A *filtered Lie algebra* is a Lie algebra L which is also a (decreasing) filtered module, such that (i) $[F^iL, F^jL] \subset F^{i+j}L$ (i.e. the bracket $L \otimes L \to L$ is filtration preserving), and (ii) $F^1L = L$.

A *complete Lie algebra* is a filtered Lie algebra which is complete as a filtered module. With the evident notion of morphism these form a category $\langle\mathrm{CLA}\rangle$.

As in (3.21) there is defined a *completion* of filtered LAs. The associated graded module $\mathrm{gr}(L)$ of a filtered LA is a Lie algebra (with internal grading) satisfying $\mathrm{gr}^0 L = 0$.

Universal enveloping (3.48)

Let L be a LA. We consider the universal enveloping algebra $j : L \to UL$. A filtration of L yields an *associated* filtration of UL. By definition this is the minimal filtration of UL (as AA), such that j is a filtered map. Explicitly, F^mUL $(m \geqslant 1)$ is the submodule generated by all elements $j(x_1)\dots j(x_k)$, $x_i \in F^{m_i}L$, $\sum m_i \geqslant m$ and m_i, $k \geqslant 1$.

We can also describe this associated filtration as being produced by the natural surjection $\mathsf{T}L \twoheadrightarrow UL$ (with kernel J generated by $\{xy - yx - [x,y] \mid x, y \in L\}$). Here $\mathsf{T}L$ is the filtered tensor algebra, see (3.31). Then it is clear that the usual Hopf diagonal of UL is a filtered AA-map, i.e. UL is a filtered Hopf algebra.

Composition with completion of filtered AAs yields the functor $\widehat{U}$ into $\langle\mathrm{CHA}\rangle$, defined on the category of filtered LAs, and thus especially on the category of CLAs. We may describe $\widehat{U}L$ as the quotient of $\widehat{\mathsf{T}}L$ by the complete submodule $\widehat{J}$. (Recall that completion preserves exactness (C.16), hence $\widehat{J}$ is an ideal). We have adjoint functors

$$\text{(3.49)} \qquad \langle\mathrm{CHA}\rangle \underset{\mathcal{P}}{\overset{\widehat{U}}{\leftrightarrows}} \langle\mathrm{CLA}\rangle$$

where $\mathcal{P}A$ is the LA of primitives (3.41) endowed with the induced filtration. In other words, $\mathcal{P}A$ is the kernel in $\langle\mathrm{CLA}\rangle$ of the reduced diagonal $\overline{\Delta} : \overline{A} \to \overline{A} \mathbin{\widehat{\otimes}} \overline{A}$.

Remarks. We re-obtain the functor $\widehat{U}$ of (3.40): If L is a LA, the augmental filtration of UL is associated with the lower central series. In fact, $j(\Gamma^n L) \subset \overline{UL}^{\,n}$ and the augmental filtration is minimal. By adjoint functor arguments, we have a canonical isomorphism $\widehat{U}L = \widehat{U}\widehat{L}$.

Symmetric algebra (3.50)

Let W be a complete module with $\mathrm{gr}^0 W = 0$. The universal CHA of W (endowed with the trivial bracket) is called the *complete symmetric algebra* over V, denoted by $\widehat{\mathsf{S}}W$. It is the completion of the symmetric algebra $\mathsf{S}W$. Hence $\widehat{\mathsf{S}}W = \widehat{\bigoplus}_{n\geqslant 0}\widehat{\mathsf{S}}^n W$ and we have, for all n, a short exact (cf. (C.16)) sequence of complete modules

$$(*) \qquad \widehat{J}^n \hookrightarrow W^{\widehat{\otimes} n} \overset{q}{\twoheadrightarrow} \widehat{\mathsf{S}}^n W$$

where $J^n \subset W^{\otimes n}$ is the module generated by $\{z - \pi(z) \mid z \in W^{\otimes n}, \pi \in S_n\}$, endowed with the induced filtration. The symmetric group S_n operates on n-fold (filtered, or complete) tensor product by permuting the factors.

Taking the complete direct sum over the sequences $(*)$ for $n \geqslant 0$, we get the short exact sequence $\widehat{J} \hookrightarrow \widehat{\mathsf{T}}W \twoheadrightarrow \widehat{\mathsf{S}}W$ in $\langle\mathrm{cpl}\rangle$. It follows from $W = F^1 W$ that these complete direct sums are ordinary products (see (C.7)), in particular $\widehat{\mathsf{S}}W = \prod_{n\geqslant 0}\widehat{\mathsf{S}}^n W$ (product filtration).

The map $\sigma^n : W^{\widehat{\otimes} n} \longrightarrow W^{\widehat{\otimes} n}$, $z \mapsto \sum_{\pi\in S_n}\pi(z)$ is zero on $\widehat{J}^n$ and thus factors through q in $(*)$.

PROPOSITION (3.51)

The induced map $s^n : \widehat{\mathsf{S}}^n W \longrightarrow W^{\widehat{\otimes} n}$ *is normal injective and identifies* $\widehat{\mathsf{S}}^n W$ *with the symmetric tensors* $\{z \mid \pi(z) = z, \pi \in S_n\}$.

The map $e : \widehat{\mathsf{S}}W \longrightarrow \widehat{\mathsf{T}}W$ *given by* $e^n = s^n/n!$ *is a filtered section for the canonical quotient map* q.

Proof. In fact, $q \circ s^n = n!$ which is an isomorphism, because $W^{\widehat{\otimes} n}$ (thus $\widehat{\mathsf{S}}^n W$) is equal to its n-th filtration, and so is a K_n-module. Hence s^n has a filtered retraction and is thus a normal injection. The last assertion is now also obvious. It is clear that $\mathrm{Im}(\sigma^n) = \mathrm{Im}(s^n)$ consists of symmetric tensors, and if z is such one then $\sigma^n(z/n!) = \sigma^n(z)/n! = z$ proving the second assertion. □

Free complete Lie algebra (3.52)

Let V be a free filtered module, see (3.31). Then the free Lie algebra $\mathsf{L}V$ over (the free module) V has the following associated filtration: $F^n\mathsf{L}V$ is the module generated by brackets (of any finite order) of elements x_i such that there are integers $m_i \geqslant 0$ with $x_i \in F^{m_i}V$ and $\sum m_i \geqslant n$. Endowed with this filtration $\mathsf{L}V$ is called the *free* filtered LA over V.—Recall that $\mathsf{L}V = \bigoplus_{n=1}^{\infty}\mathsf{L}^n V$ where $\mathsf{L}^n V$ is the module generated by the n-th order brackets. It is clear that this is a direct sum decomposition of filtered modules.

Let $\widehat{\mathsf{L}}V$ be the completion. It is called the *free* CLA over V. The natural filtered map $V \to \widehat{\mathsf{L}}V$ has the obvious universal property. We have $\widehat{\mathsf{L}}V = \prod_{n=1}^{\infty}\widehat{\mathsf{L}}^n V$, since $\mathsf{L}^n V$ and so its completion coincides with its n-th filtration.

THEOREM (3.53)

The filtered map

$$(*) \qquad \rho : \widehat{\mathsf{T}}V \to \widehat{\mathsf{L}}V, \qquad x_1 \ldots x_n \mapsto \begin{cases} \frac{1}{n}[x_1,[x_2,\ldots[x_{n-1},x_n]\ldots]] & n > 0 \\ 0 & n = 0 \end{cases}$$

where $x_i \in V$, *is left inverse to the obvious filtered map* $j : \widehat{\mathsf{L}}V \to \widehat{\mathsf{T}}V$.

The latter is thus a normal injection, identifying $\widehat{\mathsf{L}}V$ with the sub-CLA of $\widehat{\mathsf{T}}V$ generated by V. Notice that we actually define ρ on $\mathsf{T}V$ and then use the universal property of the completion.

Proof. Let $\rho' : \mathsf{T}V \to \mathsf{L}V$ be given by the formula $(*)$ but with the factor $1/n$ omitted. It is classical that $j : \mathsf{L}V \to \mathsf{T}V$ followed by ρ' is multiplication with n on $\mathsf{L}^n V$; cf. [Q], [Bki]. It is obvious that these are filtered maps, hence they induce maps on the completions. Since the complete module $\widehat{\mathsf{L}^n}V$ coincides with its n-th filtration, multiplication with n is an automorphism. This shows that ρ is well-defined and the desired property follows. □

PBW–Theorem (for CLAs) (3.54)
For $M \in \langle \mathrm{CLA} \rangle$ *the natural map*

$$e : \widehat{\mathsf{S}}M \to \widehat{U}M, \qquad x_1 \dots x_n \mapsto \frac{1}{n!} \sum_{\sigma \in S_n} j(x_{\sigma(1)}) \dots j(x_{\sigma(n)})$$

where $x_i \in M$, *is a filtered isomorphism compatible with the diagonals.*

In particular, $j : M \to \widehat{U}M$ has a natural filtered retraction.

Proof. The sum (in the definition of e) lies in filtration n and may thus be divided uniquely by $n!$ in *this* filtration (though possibly non-uniquely in $\widehat{U}M$). Thus e is well-defined. In case of a free CLA it is fairly easy to reduce to the graded analog of the theorem, see below. In the general case, we construct a diagram

$$(*) \qquad L_1 \underset{d_0}{\overset{d_1}{\rightrightarrows}} L_0 \overset{p}{\twoheadrightarrow} M \qquad \text{in } \langle \mathrm{CLA} \rangle$$
(with $s : L_0 \to L_1$)

with L_0, L_1 free and $d_0 s = 1 = d_1 s$, $pd_0 = pd_1$; moreover p is normal surjective and $d_0(\operatorname{Ker} d_1) = \operatorname{Ker} p$ as submodules of L_0. It is then immediate that p is a coequalizer of d_0, d_1 in $\langle \mathrm{cpl} \rangle$ and hence in $\langle \mathrm{CLA} \rangle$.

For bringing this diagram into existence one may proceed as follows. First construct a normal surjection from a free filtered module to M by induction on the filtration, the extension to the free CLA is p. Then, by adding more generators to L_0, factorize the map $(1,1) : L_0 \to L_0 \times_M L_0$ into $d \circ s$ where $d = (d_0, d_1)$ is surjective and its source L_1 is a free CLA.

Now left adjoint functors preserve coequalizer diagrams, hence applying $\widehat{U}$ to $(*)$ yields a coequalizer diagram

$$(**) \qquad A_1 \underset{d_0}{\overset{d_1}{\rightrightarrows}} A_0 \overset{p}{\twoheadrightarrow} A \qquad \text{in } \langle \mathrm{CAA} \rangle$$
(with $s : A_0 \to A_1$)

with a map s (short for $\widehat{U}s$ etc.) satisfying $d_0 s = 1 = d_1 s$. It follows that d_0 is surjective, hence $J = d_0(\operatorname{Ker} d_1)$ is an ideal of A_0. An easy computation shows that p is universal among all CAA-maps annihilating J. Hence $A = A_0/\widehat{J}$ which implies that $(**)$ is a coequalizer also in $\langle \mathrm{cpl} \rangle$.

The same arguments show that S applied to $(*)$ yields a coequalizer diagram in $\langle \mathrm{cCAA} \rangle$ and *a posteriori* in $\langle \mathrm{cpl} \rangle$. Since e is natural and a filtered isomorphism for L_0, L_1 the assertion follows. $(\Box)$

Proof of the PBW–Theorem for any free CLA M.

Here we apply the method of "reduction to associated graded objects", which in the present context is discussed below. Consider diagram (3.56) below in which the horizontal maps are surjective, and the left vertical map e is surjective by the "graded PBW–Theorem" (3.60) proved below. Therefore, in order to show that all maps are isomorphisms (then we are done) it suffices to show that the composite is injective. By a standard argument (see proof of (3.60)) this reduces to show that the restriction $\mathrm{gr}(M) \to \mathrm{gr}(\widehat{U}M)$ is injective. Disregarding differentials, this is (for some free filtered module V) the associated graded of the filtered map $j : \widehat{\mathsf{L}}V \to \widehat{\mathsf{T}}V$ which has a filtered retraction by (3.53). This completes the proof. $\Box$

COROLLARY (3.55)

For any $M \in \langle\mathrm{CLA}\rangle$ the horizontal maps in the obvious diagram below are surjective. These are isomorphisms if $\mathrm{gr}(M)$ is flat.

(3.56)
$$\begin{array}{ccc} \widehat{\mathsf{S}}\,\mathrm{gr}(M) & \longrightarrow & \mathrm{gr}(\widehat{\mathsf{S}}M) \\ {\scriptstyle e}\downarrow & & \downarrow{\scriptstyle \mathrm{gr}(e)} \\ \widehat{U}\,\mathrm{gr}(M) & \longrightarrow & \mathrm{gr}(\widehat{U}M) \end{array}$$

Proof. The horizontal maps are quotients of the canonical map $\pi : \widehat{\mathsf{T}}\mathrm{gr}(M) \to \mathrm{gr}(\widehat{\mathsf{T}}M)$. Since this map is always surjective by (3.17), the first assertion follows. For the second, we may now use that the vertical maps are bijective. The top map is a retract of π by (3.51), and the latter is an isomorphism if $\mathrm{gr}\,M$ is flat (C.24). $\Box$

Excursion: The graded case (3.57)

Recall the general procedure of how graded modules with their tensor product (additive monoidal structure) give rise to the obvious categories $\langle$graded AA$\rangle$, $\langle$graded LA$\rangle$, $\langle$graded HA$\rangle$. According to the filtered situation we assume graded "algebras" to be *reduced*, i.e. trivial in degree 0. (Here we use the term "algebra" to denote any of the mentioned species' of algebraic objects arising in presence of an additive monoidal category).

Analogously, using *complete* graded modules with their complete tensor product, see (3.19), we define categories $\langle$graded CAA$\rangle$, $\langle$graded CLA$\rangle$, $\langle$graded CHA$\rangle$. In the first and second case these are just the full subcategories of those objects being complete as graded module; in the third case however this is not so: the diagonal of a graded CHA A is a map into $A \mathbin{\widehat{\otimes}} A$. The completion functor lifts (specializes) to all three cases in a canonical way.

In order to justify the notation, notice that one may regard $\langle$graded CHA$\rangle$ *etc.* as the category of "CHAs with splittings" *etc.*, see (3.20).

Reduction to the graded case (3.58)

The functor 'gr' from filtered to graded modules lifts (specializes) to functors from *filtered* to *graded* "algebras".

There is however a restriction resulting from the fact that 'gr' is *not* monoidal (B.2) in general: the associated graded of a filtered coalgebra is *not* a graded coalgebra in general; however this holds on certain conditions, cf. (C.24).

We may thus consider 'gr' as a functor from ⟨CAA⟩ to ⟨graded CAA⟩, from ⟨CLA⟩ to ⟨graded CLA⟩, and (suitably restricted) from ⟨CHA⟩ to ⟨graded CHA⟩.

The functors $\widehat{\mathsf{T}}$, $\widehat{U}$ and $\widehat{\mathsf{S}}$ (graded case) (3.59)

First let L be a Lie algebra over K. Let UL be the universal enveloping algebra of L, and let $\mathsf{S}L$ resp. $\mathsf{T}L$ be the symmetric resp. tensor algebra over the module L. Now let L be endowed with a grading (compatible with the bracket). As is well-known we get an induced grading of $\mathsf{T}L$, $\mathsf{S}L$ and UL. So these are graded augmented algebras each satisfying the obvious universal property. Applying the completion functor $(- \otimes \mathsf{K}_*)$ yields graded CAAs (in fact graded CHAs) $\widehat{\mathsf{T}}L$, $\widehat{\mathsf{S}}L$ and $\widehat{U}L$. This is of particular interest when L is a graded CLA.

PBW–THEOREM (FOR GRADED CLAS) (3.60)

For any graded CLA M, the natural graded map $e : \widehat{\mathsf{S}}M \to \widehat{U}M$ is bijective and compatible with the canonical diagonals.

Proof. We first show that e is surjective. Given any ordinary K-Lie algebra L, we consider the standard (increasing) filtration of UL resp. $\mathsf{S}L$ where F_n is generated by the k-fold products, $k \leqslant n$, of elements of $\mathrm{Im}(L)$. By well-known arguments the associated graded of UL is a commutative algebra generated by $\mathrm{gr}_1(UL)$. The canonical map $\mathsf{S}L \to \mathrm{gr}_* UL$ is thus surjective. Consider this map in degrees $\leqslant n$ and tensorized with K_n; we get a surjection which identifies with the associated graded of the filtered map $e_n : F_n(\mathsf{S}L) \otimes \mathsf{K}_n \to F_n(UL) \otimes \mathsf{K}_n$ which is (well-)defined by the same formula as e is. We conclude that e_n is surjective.

Now if M is a graded CLA (considered as an ordinary Lie algebra, as in (3.20)), then UM, $F_n UM$ are graded modules which coincide in degrees $\leqslant n$ since $M^0 = 0$. The graded maps $e : \widehat{\mathsf{S}}M \to \widehat{U}M$ and e_n coincide in degree n. Since n is arbitrary we conclude that e is surjective.

The next step is the injectivity of e in case M free. The only hint that remains to be given is the following. In order to show that a map $f : \widehat{\mathsf{S}}M \to E \in$ ⟨graded CCC⟩ is injective, it suffices that its restriction to M is injective. The (standard) proof is by induction using that $\widehat{\mathsf{S}}M$ is connected (in the sense of (2.13)) and $\widehat{\mathsf{S}}M$ and E are flat. — Here "CCC" means complete coaugmented coalgebra, i.e. ⟨graded CCC⟩ is the category of coaugmented coalgebras over $\mathcal{V} =$ ⟨complete graded modules⟩ (with the complete tensor product).

The rest of the proof is analogous to (and, in fact, follows from) the filtered case. □

N° 3 Normal complete Hopf algebras

We show first that ⟨CLA⟩ is canonically equivalent to the full subcategory of ⟨CHA⟩ consisting of all objects of the form $\widehat{U}M$ (up to isomorphism). It turns out that these are precisely the *normal* CHAs. Then we prove that a CHA is normal if its associated graded module is flat (which is always the case if K is a field).

We call a CHA A *normal* if its reduced diagonal

$$\overline{\Delta} : \overline{A} \to \overline{A}^{\widehat{\otimes} 2}$$

is a normal map. For any CHA A, we define a natural filtered map

$$\overline{\overline{\Delta}} = (1-\tau, \overline{\Delta}\otimes 1 - 1\otimes\overline{\Delta}) : \overline{A}^{\widehat{\otimes}2} \to \overline{A}^{\widehat{\otimes}2} \oplus \overline{A}^{\widehat{\otimes}3}.$$

For any $M \in \langle\mathrm{CLA}\rangle$, we consider the natural sequence of solid arrows in the diagram below, where $A = \widehat{U}M$.

We call a sequence of filtered maps *normal exact* if it is exact as a sequence of module maps and all maps are normal, cf. (C.13).

$$(3.61)\qquad 0 \to M \xrightarrow{\ j\ } \overline{A} \underset{\mu}{\overset{\overline{\Delta}}{\rightleftarrows}} \overline{A}^{\widehat{\otimes}2} \underset{\lambda}{\overset{\overline{\overline{\Delta}}}{\rightleftarrows}} \overline{A}^{\widehat{\otimes}2} \oplus \overline{A}^{\widehat{\otimes}3}$$

THEOREM (3.62)

For any CLA M, the CHA $\widehat{U}M = A$ is normal.

The sequence of solid arrows is normal exact, and there are (natural) filtered maps μ, λ which are splittings in the sense that $\overline{\Delta}\mu + \lambda\overline{\overline{\Delta}} = 1$.

Proof. It is clear that consecutive maps compose to zero. Since $e : \widehat{\mathsf{S}}M \to \widehat{U}M$ is compatible with the diagonals, we are reduced to the case where M is abelian. Thus consider the diagram with $A = \widehat{\mathsf{S}}M$. The latter has a canonical complete direct sum decomposition and we may define the maps μ, λ as in the proof of (D.20). They satisfy $\mu\overline{\Delta} = 1$ on $\widehat{\bigoplus}_{k\geqslant 2}\widehat{\mathsf{S}}^k M$, and $\overline{\Delta}\mu + \lambda\overline{\overline{\Delta}} = 1$. The exactness (in the ordinary sense) follows. Moreover we obtain $\overline{\Delta}\,\mu\,\overline{\Delta} = \overline{\Delta}$ and $\overline{\overline{\Delta}}\,\lambda\,\overline{\overline{\Delta}} = \overline{\overline{\Delta}}$, hence $\overline{\Delta}$ and $\overline{\overline{\Delta}}$ are normal, and j is normal because it admits a filtered module retraction; cf. (C.10). □

COROLLARY (3.63)

For $M \in \langle\mathrm{CLA}\rangle$ the canonical map $M \to \mathcal{P}\widehat{U}M$ is a filtered isomorphism. □

THEOREM (3.64)

Let $A \in \langle\mathrm{CHA}\rangle$ be normal. Then the canonical map $\widehat{U}\mathcal{P}A \to A$ is a filtered isomorphism.

Proof. We prove by induction that $\alpha : \widehat{U}\mathcal{P}A \to A$ is a filtered isomorphism *mod* F^{m+1}, which means that $\mathrm{gr}(\alpha)^{\leqslant m}$ is an isomorphism. This is clear for $m = 0$, hence suppose α is an isomorphism *mod* F^m. Consider the diagram

$$(*)\qquad \begin{array}{ccccccc}
\mathcal{P}B & \hookrightarrow & \overline{B} & \xrightarrow{\overline{\psi}} & \overline{B}^{\widehat{\otimes}2} & \xrightarrow{\overline{\overline{\psi}}} & \overline{B}^{\widehat{\otimes}2} \oplus \overline{B}^{\widehat{\otimes}3} \\
\Big\downarrow{=} & & \Big\downarrow{\alpha} & & \Big\downarrow{\alpha^{\widehat{\otimes}2}} & & \Big\downarrow{\alpha^{\widehat{\otimes}2}\oplus\alpha^{\widehat{\otimes}3}} \\
\mathcal{P}A & \hookrightarrow & \overline{A} & \xrightarrow{\overline{\Delta}} & \overline{A}^{\widehat{\otimes}2} & \xrightarrow{\overline{\overline{\Delta}}} & \overline{A}^{\widehat{\otimes}2} \oplus \widehat{A}^{\widehat{\otimes}3}
\end{array}$$

where $B = \widehat{U}\mathcal{P}A$ and ψ its diagonal. The top row of the diagram is normal exact, and the left vertical map a filtered isomorphism by (3.63). Since $\overline{\Delta}$ is normal by assumption, the sequence $\mathcal{P}A \hookrightarrow \overline{A} \to \overline{A}^{\widehat{\otimes}2}$ is normal exact. We know that 'gr' carries a normal exact sequence to an exact sequence, and a normal injection to an injection, see (C.15), (C.11).

Since we consider here the augmentation ideals it follows from the inductive hypothesis, (C.20) and (3.18) that $\alpha^{\widehat{\otimes}2}$ and $\alpha^{\widehat{\otimes}2}\oplus\alpha^{\widehat{\otimes}3}$ are filtered isomorphisms *mod* F^{m+1}. In other words, $\mathrm{gr}^{\leqslant m}$ carries $(*)$ to a diagram where the two right vertical maps are isomorphic, and clearly so is the left. Hence the 5-lemma applied to the diagram of associated graded maps yields that $\mathrm{gr}(\alpha)^{\leqslant m}$ is an isomorphism. This completes the induction, and the proof of the theorem. □

Together with the PBW–Theorem we obtain the following "adjoint PBW–Theorem" for CHAs.

COROLLARY (3.65)

For any normal CHA A, there is a canonical filtered module isomorphism $\widehat{\mathsf{S}}\mathcal{P}A \xrightarrow{\approx} A$ *compatible with the canonical diagonals.* Hence:

$\mathcal{P}A$ is a filtered module retract of A in a natural way.

A map $f : A \to A'$ between CHAs as above is a filtered isomorphism iff $\mathcal{P}f$ is a filtered isomorphism. □

We also conclude: the canonical functors are adjoint equivalences between ⟨CLA⟩ and ⟨normal CHA⟩; for variants see N°4. Of course there are analogous results for graded CHAs where the condition of normality is vacuous, cf. (3.57).

The following result shows (in particular) that over a field, *all* CHAs are normal. In general, it is an open question whether there are non-normal CHAs. However we do not need this here, since all CHAs of interest "come from" (complete) Lie algebras or groups, and thus *are* normal; cf. (3.91) (a).

THEOREM (3.66)

Let $A \in \langle\mathrm{CHA}\rangle$ be such that $\mathrm{gr}(A)$ *is flat. Then A is normal, and there is a canonical isomorphism* $\mathrm{gr}(\mathcal{P}A) = \mathcal{P}\mathrm{gr}(A)$.

Proof. Recall that the flatness assumption implies that $\mathrm{gr}(A)$ is a graded CHA (3.57) with diagonal $\delta = \mathrm{gr}(\Delta)$. Therefore the second assertion makes sense, claiming that

$$\mathrm{gr}\,(\mathcal{P}A) \hookrightarrow \mathrm{gr}\,(\overline{A}) \xrightarrow{\overline{\delta}} \mathrm{gr}(\overline{A})^{\widehat{\otimes}2}$$

is an exact sequence. This is the associated graded of the sequence

$$\mathcal{P}A \hookrightarrow \overline{A} \xrightarrow{\overline{\Delta}} \overline{A}^{\widehat{\otimes}2}$$

and (by (C.15)) it suffices to show that the latter sequence is normal exact. As it is exact and the inclusion is normal by definition, this means that A is normal.

In order to show $\overline{\Delta}$ normal, we prove that the sequence

$$(*) \qquad \overline{A} \xrightarrow{\overline{\Delta}} \overline{A}^{\widehat{\otimes}2} \xrightarrow{\overline{\overline{\Delta}}} \overline{A}^{\widehat{\otimes}2} \oplus \overline{A}^{\widehat{\otimes}3}$$

is normal exact. By (C.15) this follows once we know that $(*)$ composes to zero and yields via 'gr' an exact sequence. The former is obvious, and the latter a consequence of the graded analogs of (3.64) and (3.62). Notice that 'gr' applied to $(*)$ yields the analogous sequence for the graded CHA gr A. □

Remark. What we actually need here are the canonical isomorphisms $\mathrm{gr}(A)^{\widehat{\otimes} n} = \mathrm{gr}(A^{\widehat{\otimes} n})$ for $n = 2, 3$. These hold if $\mathrm{gr}(A)$ is flat or if A is splittable, see (C.24).

N° 4 Rigidity

Recall the canonical equivalence between $\langle \mathrm{CLA} \rangle$ and the category $\langle$normal CHA$\rangle$. Our main goal is to prove that it restricts to the full subcategories of rigid objects.

In fact, there is a notion of *rigidity* of CLAs analogous to CAAs; see (3.32), or the proof below. The results proved in § 1, N° 4 hold without change in proof; one only has to replace $\widehat{\mathsf{T}}$ by $\widehat{\mathsf{L}}$.

A CHA is said to be *rigid*, if it is so-called as CAA.

LEMMA (3.67)

(a) *Let $M \in \langle \mathrm{CLA} \rangle$. Then $\mathrm{gr}(M)$ is flat iff $\mathrm{gr}(\widehat{U}M)$ is flat. Further M is rigid iff $\widehat{U}M$ is rigid.*

(b) *Let $A \in \langle \mathrm{CHA} \rangle$ be normal. Then $\mathrm{gr}(A)$ is flat iff $\mathrm{gr}(\mathcal{P}A)$ is flat. Further A is rigid iff $\mathcal{P}A$ is rigid.*

Proof. Ad (a). When $\mathrm{gr}(M)$ is flat then clearly so is $\widehat{\mathsf{T}}\mathrm{gr}(M)$, hence so is the retract $\widehat{\mathsf{S}}\mathrm{gr}(M)$. The first assertion follows thus from (3.55) and PBW.—When $\widehat{\mathsf{L}}(\mathrm{gr}^1 M) \to \mathrm{gr}(M)$ is surjective, so is $\widehat{U}\widehat{\mathsf{L}}(\mathrm{gr}^1 M) = \widehat{\mathsf{T}}(\mathrm{gr}^1 M) \to \widehat{U}\mathrm{gr}(M)$, since $U\mathrm{gr}(M)$ is generated by $\mathrm{gr}(M)$. Using (3.55) this implies that the map $\widehat{\mathsf{T}}(\mathrm{gr}^1 M) \to \mathrm{gr}(\widehat{U}M)$ is surjective, hence $\widehat{U}M$ is rigid.—For the converse note that by (3.55) the map $\mathrm{gr}^1(M) \to \mathrm{gr}^1(\widehat{U}M)$ is surjective (in fact it is bijective). Hence $\widehat{U}M$ rigid implies $\widehat{U}\widehat{\mathsf{L}}(\mathrm{gr}^1 M) \to \mathrm{gr}(\widehat{U}M)$ surjective. It suffices to show that this map has $\widehat{\mathsf{L}}(\mathrm{gr}^1 M) \to \mathrm{gr}(M)$ as a retract. This is clear by (3.55) and PBW.

Ad (b). Follows from (a), (3.63), (3.64). □

We have thus proved several equivalences of categories

$$\mathcal{P} : \langle \mathrm{CHA} \rangle' \approx \langle \mathrm{CLA} \rangle' : \widehat{U} \tag{3.68}$$

where $\langle \ldots \rangle'$ denotes a full subcategory of $\langle \ \ldots \ \rangle$. Namely, $A \in \langle \mathrm{CHA} \rangle'$ resp. $M \in \langle \mathrm{CLA} \rangle'$ means one of the following:

(i)	A is normal	M is arbitrary
(ii)	A is normal and rigid	M is rigid
(iii)	$\mathrm{gr}(A)$ is flat	$\mathrm{gr}(M)$ is flat
(iv)	$\mathrm{gr}(A)$ is flat and A rigid	$\mathrm{gr}(M)$ is flat and M rigid.

THEOREM (3.69)

The category $\langle \mathrm{CLA} \rangle'$ of rigid CLAs has arbitrary limits and colimits. Moreover, the effective epimorphisms are precisely the surjective maps, and $\widehat{\mathsf{L}}(s)$ is a projective generator.

COROLLARY (3.70)

The category $\langle \mathrm{CHA} \rangle'$ of rigid normal CHAs has arbitrary limits and colimits. Moreover, the effective epimorphisms are precisely the surjective maps, and $\mathsf{K}\langle\langle \mathbf{X} \rangle\rangle$ is a projective generator.

Proofs. The theorem is proved the same way as the corresponding result for $\langle \mathrm{CAA}\rangle'$, see (3.37). The corollary follows then from the equivalence of categories (3.68) (ii), and the isomorphism $\widehat{U}\widehat{\mathsf{L}}(s) \cong \mathsf{K}\langle\langle \mathbf{X}\rangle\rangle$, see (3.45). It only remains to verify that a map f in $\langle \mathrm{CLA}\rangle'$ is surjective iff $\widehat{U}f$ is surjective. This follows from the fact that here f surjective implies $\mathrm{gr}(f)$ surjective. □

§3 Complete Groups

In this section, K is a subring of $\mathbb{Q}$. Hence, *unspecified tensor products are over* $\mathbb{Z}$.

Our chief goal is a result of Lazard which states that the category of complete Lie algebras (over a given mild ring system) is *isomorphic* to a category of certain 'complete groups'. This is a variant of the classical Lie correspondence. The main observation is that the Hausdorff series defines, in a natural way, a group law on any complete Lie algebra. Then the results of the previous section yield analogous results on the relationship between (complete) groups and complete Hopf algebras.

N° 1 Nilpotent groups

We need certain functors Π^{-1} defined on the category of nilpotent groups $\langle \mathrm{nGp}\rangle$, for any set Π of primes. These behave very much like (and extend) the ordinary tensor product $-\otimes \Pi^{-1}\mathbb{Z}$ on the category of abelian groups, which is discussed in §1, N° 1.

For a group G let $G = \Gamma^1 G \supset \Gamma^2 G \supset \dots$ be the *lower central series*, $\Gamma^n G = (G, \Gamma^{n-1}G)$, the subgroup generated by all commutators $(x,y) = x^{-1}y^{-1}xy$ where $x \in G$, $y \in \Gamma^{n-1}G$. These are normal subgroups, since we have the relation $(\Gamma^i G, \Gamma^j G) \subset \Gamma^{i+j}G$ for all i,j. Among all filtrations of G satisfying this relation, the lower central series is *minimal*.

Associated graded LA of a filtered group (3.71)

If a descending filtration $G = F^1G \supset F^2G \supset \dots$ of a group G satisfies $(F^iH, F^jH) \subset F^{i+j}H$, then for all n the canonical extension (= short exact sequence) $\mathrm{gr}^n G \hookrightarrow G/F^{n+1}G \twoheadrightarrow G/F^nG$ is central, where $\mathrm{gr}^n G = F^nG/F^{n+1}G$. Moreover, $\mathrm{gr}\, G$ is a graded LA over $\mathbb{Z}$, where bracket is induced by commutator; cf. [Se].

Now let G be filtered by the lower central series. Then $\mathrm{gr}^1 G = G_{\mathrm{ab}}$ is the *abelianization* of G. It is not difficult to see that $\mathrm{gr}\, G$ is generated, as LA, by $\mathrm{gr}^1 G$; cf. [Se]. Recall that G is *nilpotent* (of class $\leqslant r$) if $\Gamma^{r+1}G = 1$ for some r.

Definitions (3.72)

Let Π be a set of primes. An element of the multiplicative subset of $\mathbb{Z}$ generated by $\{1\} \cup \Pi$ is said to be a Π-*integer*. These are precisely the positive integers invertible in the ring $\Pi^{-1}\mathbb{Z}$.

Let G be a group. An element $x \in G$ is called a Π-*torsion* element if $x^a = e$ for some Π-integer a. Thus G is said to be Π-*torsion* (resp. Π-*torsion-free*) if the subset of Π-torsion elements is G (resp. $\{e\}$). Further G is said to be Π-*radicable* (resp. Π-*uniquely-radicable*) if, for every Π-integer a, the mapping $x \mapsto x^a$ is surjective (resp. bijective). — We say also 'divisible' for 'radicable' in the abelian case.

The category of Π-uniquely-radicable nGps is denoted by $\Pi^{-1}\langle\text{nGp}\rangle$.

Proposition (3.73)

(a) *All groups in a short exact sequence* $H \hookrightarrow G \twoheadrightarrow G'$ *are* Π*-uniquely-radicable if one of the following holds:* (i) H, G *are* Π*-uniquely-radicable and* G' *is nilpotent;* (ii) G, G' *are* Π*-uniquely-radicable;* (iii) H, G' *are* Π*-uniquely-radicable and* H *is in the center of* G.

(b) *The class of* Π*-uniquely-radicable groups is closed under limits in* $\langle\text{Gp}\rangle$.

Proof. The only difficulty is part (i) of (a) (which is trivial when G' is abelian). Here one uses the fact ([Laz | p. 159]) that in any nilpotent group N one has

$$(x^a, y) = 1 \Rightarrow (x, y)^{a^{r-1}} = 1 \tag{3.74}$$

where r = class of N. Hence, when N is Π-torsion-free (e.g., Π-uniquely-radicable), then $(x^a, y^b) = 1$ with Π-integers a, b implies $(x, y) = 1$. □

Corollary (3.75)

Let N *be a nilpotent group.*

(a) *If* N *is* Π*-uniquely-radicable, so is* $\Gamma^k N$ *and* (thus) $N/\Gamma^k N$ *and* $\text{gr}^k N$.

(b) N *is* Π*-uniquely-radicable iff* $\text{gr}\, N$ *is* Π*-uniquely-divisible.*

Proof. If N is Π-radicable, so is $\text{gr}^1 N$ and hence $\text{gr}\, N$. By induction on n, this implies that for any $x \in \Gamma^k N$ and Π-integer a, there is $y \in \Gamma^k N$ with $y^a \equiv x \bmod \Gamma^n N$. Hence $\Gamma^k N$ is Π-radicable, and the "uniqueness" part is trivial. The other assertions follow from this and (3.73). □

Theorem (3.76)

There is a functor $\Pi^{-1} : \langle\text{nGp}\rangle \to \Pi^{-1}\langle\text{nGp}\rangle$ *left adjoint - left inverse to the embedding.*

If f *is an injective* (resp. *surjective*) *nGp-map, so is* $\Pi^{-1}f$. *The functor* Π^{-1} *is exact. When* $H \subset N$ *is a central subgroup, so is* $\Pi^{-1}H \subset \Pi^{-1}N$.

When N *is abelian,* $\Pi^{-1}N = N \otimes \Pi^{-1}\mathbb{Z}$. □

For a proof see [War | §8], or [BouK | Ch. V, § 2].

The first assertion means that any nGp N admits a universal map $c = c_N : N \to \Pi^{-1}N$ into a Π-uniquely-radicable nGp. We fix such a map for each nilpotent group. In particular, if N is Π-uniquely-radicable then we (may) put $c = \text{id}$.

Since Π^{-1} preserves short exact sequences of groups, it preserves "reasonable" pull-back diagrams of nilpotent groups (cf. proof of (C.4)).

As an immediate consequence of the universal property we obtain, for sets of primes $\Pi, \bar{\Pi}$, the formula: $\bar{\Pi}^{-1}\Pi^{-1} = (\bar{\Pi} \cup \Pi)^{-1}$.

Proposition (3.77)

For a map $u : N \to N'$ *with* $N' \in \Pi^{-1}\langle\text{nGp}\rangle$ *are equivalent:* (i) u *is universal,* (ii) $\text{gr}(u)$ *is universal,* (iii) u *has the following properties:*

(1) $\text{Ker}(u)$ is precisely the Π-torsion of N,

(2) $x \in N' \Rightarrow x^a \in \text{Im}(u)$ for some Π-integer a.

Remarks (3.78)

(a) In particular, the set of Π-torsion elements of N, being the kernel of c_N, is a normal subgroup of N. Moreover, N is Π-torsion-free iff the mapping $x \mapsto x^a$ on N is injective for all Π-integers a.

(b) The proof reveals the canonical isomorphism $\Pi^{-1}(\Gamma^k N) \to \Gamma^k(\Pi^{-1}N)$.

(c) Condition (2) means that $\mathrm{Im}(u)$ generates N' as a Π-uniquely-radicable nGp. This is related to the following: if $N'' \subset N'$ are nGps, the Π-*envelope* $\{x \in N' | x^a \in N''$ for some Π-integer $a\}$ is a subgroup of N'' (see [War | 3.25]). We conclude that if $N' \in \Pi^{-1}\langle\mathrm{nGp}\rangle$, the inclusion of N'' into its Π-envelope is universal.

Proof. (i)⇒(iii). We have to show that $c_N : N \to \Pi^{-1}N$ has the properties (1)–(2). This is clear if N is abelian. The general case follows by induction, and diagram chase (consider the diagram formed by the k-th central extension associated with N and its image under Π^{-1}). (iii)⇒(i). We know that c_N satisfies (1)–(2). Hence, if u does the same, one easily sees that the induced map $\Pi^{-1}u : \Pi^{-1}N \to N'$ is bijective. Of course, this means that u is universal.

(i)⇒(ii). If u is universal, it is immediate that $\Pi^{-1}\mathrm{gr}^1(u) : \Pi^{-1}N_{\mathrm{ab}} \to N'_{\mathrm{ab}}$ is surjective. Hence $\Pi^{-1}\mathrm{gr}(u)$ is surjective, since $\mathrm{gr}(N')$ is generated by $\mathrm{gr}^1(N')$. Now by descending induction $\Pi^{-1}(N/\Gamma^k N) \to N'/\Gamma^k N'$ (induced by u, the latter is Π-uniquely-radicable by (3.75)) is an isomorphism. Hence $\Pi^{-1}\mathrm{gr}(u)$ is an isomorphism. (ii)⇒(i). We know that $\Pi^{-1}\mathrm{gr}(c_N)$ is an isomorphism. Hence, if $\Pi^{-1}\mathrm{gr}(u)$ is an isomorphism, so is $\mathrm{gr}(\Pi^{-1}u)$. This implies that $\Pi^{-1}u$ is an isomorphism (by induction and 5-lemma, using nilpotence). □

N° 2 Complete groups

The definition of complete groups is completely analogous to that of complete Lie algebras. In order to show that every group has a completion, we need the results of N° 1.

Notation. We call a group G a K-*group*, if it is Π-uniquely-radicable, where Π is the unique set of primes such that $\mathsf{K} = \Pi^{-1}\mathbb{Z}$. For a nilpotent group N, we use $N \otimes \mathsf{K} = \Pi^{-1}N$.

Let K_* *be a ring system.* By assumption, $\mathsf{K} = \mathsf{K}_0 \subset \mathsf{K}_1 \subset \dots$ are subrings of $\mathbb{Q}$. Recall that Π_n denotes the minimal set of primes, such that $\mathsf{K}_n = \Pi_n^{-1}\mathsf{K}$. Then, given a K-module X, to say that X is a K_n-module means that X is Π_n-uniquely-divisible; cf. §1, N° 1.

Definitions (3.79)

A *filtered group* is a group G with a sequence of subgroups $G = F^1G \supset F^2G \supset \dots$ such that $(F^iG, F^jG) \subset F^{i+j}G$. — A *filtered* K-*group* is a filtered group G such that all the subgroups F^mG are K-groups.

A *complete group* (over K_*) is a filtered K-group H, such that $\mathrm{gr}^n H$ is a K_n-module, and $H = \lim\{H/F^nH\}$. — These form a category $\langle\mathrm{CGp}\rangle$ where the morphisms are the filtered group maps.

Notice that by a *filtration* on a group G we mean a sequence of subgroups giving G the structure of a filtered group. We remark that the results of this subsection (at least those being only concerned with CGps and CLAs) hold analogously for groups resp. Lie algebras filtered by a central series.

There is an obvious notion of *rigid* CGp (compare with the CLA case): H is rigid if $\operatorname{gr} H$ is generated *as a graded CLA* by $\operatorname{gr}^1 H$. Over $\mathbb{Q}$, the rigid CGps are precisely the *Mal'cev groups* considered in [Q]. The results of § 1, N° 4 have analogs in the present context. In particular, over $\mathbb{Q}$, if a rigid CGp is *discrete* (finite filtration) then its filtration is necessarily the lower central series.

Remark (3.80)

If G is a filtered K-group, then $\operatorname{gr} G$ is a K-module. Conversely, if G is a filtered group, such that $\operatorname{gr} G$ is a K-module and $G = \lim\{G/F^n G\}$, then G is a filtered K-group. (In fact, there is an analog of (3.8) for filtered groups). Hence the condition on complete groups being filtered K-groups (and not just filtered groups) is vacuous. We impose this condition for the sake of analogy with the Lie algebra case. Moreover, there is a notion of K-group for rings that are not subrings of $\mathbb{Q}$, e.g. one can take for K any $\mathbb{Q}$-algebra or the p-adic integers (see [War | p. 83–92]). It is likely that the results of this subsection can be generalized to such rings. (The case of nilpotent K-groups, where K is a field of characteristic 0, is discussed in [War | § 12]). Of course, the notion of K-group (group with exponents in K) is the non-commutative analog of the notion of K-module.

Completion of filtered groups (3.81)

Let G be a filtered group. One verifies that $H = \lim\{(G/F^n G) \otimes \mathsf{K}\}$ with $F^n H = \operatorname{Ker}\{H \twoheadrightarrow (G/F^n G) \otimes \mathsf{K}\}$ is a filtered K-group; the relation $(F^i H, F^j H) \subset F^{i+j} H$ follows from (3.74). The obvious group map $u : G \to H$ preserves filtration and induces an isomorphism $\operatorname{gr}(G) \otimes \mathsf{K} \xrightarrow{\simeq} \operatorname{gr}(H)$. (In case the ring system is constant, u is universal among all filtered maps from G into complete groups).

One constructs similarly a *CGp-completion* of G, i.e. a universal map $c : G \to \widehat{G}$ with $\widehat{G} \in \langle \mathrm{CGp} \rangle$. Here $\widehat{G}$ consists of the *coherent sequences* in the *system* of nilpotent groups $\{N_n, \rho_{n+1}\}$ where $N_n = (G/F^{n+1}G) \otimes \mathsf{K}_n$ and $\rho_{n+1} : N_{n+1} \to N_n \otimes \mathsf{K}_{n+1}$ the obvious map; cf. (C.5). The filtration of $\widehat{G}$ is obvious.— By construction, $c : G \to \widehat{G}$ induces an isomorphism $\widehat{\operatorname{gr}}(G) \xrightarrow{\simeq} \operatorname{gr}(\widehat{G})$. If any filtered map (from G into a CGp) has this *characteristic* property (that its associated graded is a completion), then it is a CGp-completion; cf. (3.14). Using this characteristic property we see that G is complete iff c is a filtered isomorphism.

As an exercise we leave the group theoretic analogs of (3.6)–(3.11) and App. C, N° 2.

Warning! We call a filtered group map $f : G \to G'$ *normal injective* if it is injective, inducing, *and* its image is a normal subgroup (recall that inducing means $F^n G = f^{-1} F^n G'$). These are precisely the kernels in the category of filtered groups. The meaning of *normal surjective* is unchanged (f is surjective in each filtration). And a map is *normal* if it factorizes into a normal surjection followed by a normal injection (the image is thus normal).— Hence, in some cases the analogs of the mentioned results are not *verbally* the same. For instance, if $\operatorname{gr}(f)$ is injective and G separated, then f is inducing injective, but of course not necessarily normal injective.

When G is any group we consider the CGp-completion $\widehat{G}$, where we consider G as a filtered group by means of the lower central series. This yields a functor

(3.82) $$\widehat{\ } : \langle \mathrm{Gp} \rangle \to \langle \mathrm{CGp} \rangle$$

left adjoint (but *not* left inverse) to the forgetful functor ϕ. Notice that if G is any group, $\widehat{G}$ is rigid. If G is nilpotent, then $\widehat{G}$ is discrete.

Recall that we have a functor $\widehat{\mathsf{K}} : \langle \mathrm{Gp} \rangle \to \langle \mathrm{CHA} \rangle$, left adjoint to the functor $\mathcal{G}$. We show that there are also canonical adjoint functors

$$\langle \mathrm{CGp} \rangle \underset{\mathcal{G}}{\overset{\widehat{\mathsf{K}}}{\rightleftarrows}} \langle \mathrm{CHA} \rangle . \tag{3.83}$$

Compatible filtrations (3.84)

Let G be a group, $j : G \to \mathsf{K}G$ the function $x \mapsto x - 1$. A group filtration of G, and an AA-filtration of $\mathsf{K}G$, are *compatible* if $j(F^m G) \subset F^m \mathsf{K}G$, $m \geqslant 1$. In any AA we have the equations

$$\begin{aligned} x^{-1}y^{-1}xy - 1 &= (yx)^{-1}(xy - yx) \\ xy - yx &= (x-1)(y-1) - (y-1)(x-1) \\ (yx)^{-1} &= 1 - (yx - 1) + (yx)^{-1}(yx-1)^2 \\ xy - 1 &= (x-1)(y-1) + (x-1) + (y-1) \end{aligned}$$

if x, y are invertible elements. Hence j induces a map $\mathrm{gr}(j) : \mathrm{gr}(G) \to \mathrm{gr}(\mathsf{K}G)$ compatible with abelian group structure and bracket (induced by commutator). Thus

$$\mathrm{gr}(j) \otimes \mathsf{K} : \mathrm{gr}(G) \otimes \mathsf{K} \to \mathrm{gr}(\mathsf{K}G)$$

is a graded LA-map, which is clearly surjective in degree 1. Recall that if $\mathrm{gr}(\mathsf{K}G)$ is flat, then it is a graded HA. Since any $x \in G$ is grouplike, we have

$$\Delta(x-1) = x \otimes x - 1 \otimes 1 = (x-1) \otimes 1 + 1 \otimes (x-1) + (x-1) \otimes (x-1) .$$

which shows that $\mathrm{Im}(\mathrm{gr}(j) \otimes \mathsf{K}) \subset \mathcal{P}\mathrm{gr}(\mathsf{K}G)$.

Given a filtration of $\mathsf{K}G$, there is clearly a maximal compatible filtration of G, the filtration *induced* by the filtration of $\mathsf{K}G$. It is given by $F^m G = G \cap (1 + F^m \mathsf{K}G)$. If j is inducing, then $\mathrm{gr}(j)$ is injective.

Given a filtered group G, there is clearly a minimal compatible filtration of $\mathsf{K}G$, the filtration *associated* with the filtration of G. Explicitly, $F^m \mathsf{K}G$ $(m \geqslant 1)$ is the sub K-module generated by all elements $(x_1 - 1) \ldots (x_k - 1)$, $x_i \in F^{m_i} G$, $\sum m_i \geqslant m$ and m_i, $k \geqslant 1$. One checks that in this situation $\mathrm{Im}(\mathrm{gr}(j) \otimes \mathsf{K})$ generates $\mathrm{gr}(\mathsf{K}G)$ as an AA.

An easy computation shows that on $\mathsf{K}(G \times G') = \mathsf{K}G \otimes \mathsf{K}G'$ the filtration associated with $F^m(G \times G') = F^m G \times F^m G'$ coincides with the tensor product of the associated filtrations. Hence $\mathsf{K}G$ with the associated filtration is a filtered Hopf algebra.

We obtain the desired functor $\widehat{\mathsf{K}}$ in (3.83) as follows: $\widehat{\mathsf{K}}H$ is the completion of the filtered Hopf algebra $\mathsf{K}H$, where the latter is endowed with the associated filtration.

For instance, when G is a group, the filtration of $\mathsf{K}G$ associated with the lower central series, is the augmental filtration. In fact, the latter is the minimal AA-filtration, and compatibility follows from the above equations (the first and third). Hence, in this case $\widehat{\mathsf{K}}G$ is the CHA considered in §2, N° 1.

The analogous construction for CLAs is discussed in (3.48). Recall that for any CHA A, $\mathcal{P}A$ has a natural CLA structure, cf. (3.49).

THEOREM (3.85)

Let $A \in \langle\text{CHA}\rangle$. *Define a filtration on* $\mathcal{G}A$ *by* $F^n\mathcal{G}A = \mathcal{G}A \cap (1 + F^nA)$, *then* $\mathcal{G}A \in \langle\text{CGp}\rangle$. *The mapping* $\exp : \mathcal{P}A \to \mathcal{G}A$ *is a filtered bijection, it induces an isomorphism*

$$(*) \qquad \mathrm{gr}(\mathcal{P}A) \xrightarrow{\simeq} \mathrm{gr}(\mathcal{G}A)$$

of graded LAs over $\mathbb{Z}$.

Proof. Using the equations in (3.84) one checks that $\mathcal{G}A$ is a filtered group. The relation $\mathcal{G}A = \lim\{\mathcal{G}A/F^k\mathcal{G}A\}$ presents no difficulty. The last assertion of the theorem implies that $\mathrm{gr}^n(\mathcal{G}A)$ is a K_n-module, hence $\mathcal{G}A$ is a CGp, cf. (3.80). It is clear that exp and log preserve filtration, hence exp is a filtered bijection, cf. (3.43). It remains to show that exp induces an isomorphism $(*)$.

For any CAA A we have the canonical isomorphism of graded LAs

$$(**) \qquad \mathrm{gr}(\overline{A}) \xrightarrow{\simeq} \mathrm{gr}(1 + \overline{A})$$

where $\overline{A}$ is regarded as sub-LA of A with the induced filtration, and $1 + \overline{A}$ is a subgroup of the invertible elements of A, with $F^n(1 + \overline{A}) = 1 + F^nA$, $n \geqslant 1$. This isomorphism $(**)$ is induced by $x \mapsto 1 + x$. It is a graded LA-map by computation, using the equations in (3.84). It is also induced by exp, since $\exp(x) = (1 + x)(1 - x + x^2 - \dots)\exp(x) = (1 + x)y$ where $y = 1 + x^2/2 + \dots \in 1 + F^{n+1}A$ if $x \in F^nA$. Hence if $x = x' + u$ with $u \in F^{n+1}A$, then $\exp(x) = \exp(x')z$ with $z \in F^{n+1}A$.

Now let A be a CHA. If in the last relation x, x' are primitive, it follows that z is grouplike. Hence exp induces a well-defined mapping $(*)$. The latter throws canonical injective graded LA-maps to $(**)$. Hence, as $(**)$ is a graded LA-map, so is $(*)$. □

This gives the desired functor $\mathcal{G}$ in (3.83), and one checks that these functors are adjoint.

N° 3 The Lazard-Mal'cev correspondence

There is a canonical CHA-isomorphism $\mathsf{K}\langle\langle\mathbf{X},\mathbf{Y}\rangle\rangle = \widehat{U}\widehat{\mathsf{L}}(\mathbf{X},\mathbf{Y})$. This follows from (3.45) and the canonical isomorphism $\widehat{U}L = \widehat{U}\widehat{L}$ for any LA L, which is a consequence of the universal property of completion (resp. adjoint functor arguments). Recall that $\widehat{U}L$ is the completion of UL endowed with the augmental filtration, while $\widehat{U}\widehat{L}$ is the completion of $U\widehat{L}$ endowed with the filtration associated with that of $\widehat{L}$.

Hence we may identify $\mathcal{P}\mathsf{K}\langle\langle\mathbf{X},\mathbf{Y}\rangle\rangle = \widehat{\mathsf{L}}(\mathbf{X},\mathbf{Y})$, cf. (3.63).

The Hausdorff group of a CLA (3.86)

We define an element of $\mathsf{K}\langle\langle\mathbf{X},\mathbf{Y}\rangle\rangle$, the *Hausdorff series*

$$\mathbf{h}(\mathbf{X},\mathbf{Y}) = \log(\exp(\mathbf{X})\exp(\mathbf{Y})) = \mathbf{X} + \mathbf{Y} + \frac{1}{2}[\mathbf{X},\mathbf{Y}] + \dots$$

By definition, this is a primitive, and thus an element of $\widehat{\mathsf{L}}(\mathbf{X},\mathbf{Y})$, see above.

Let M be any CLA. Then $M = \mathcal{P}\widehat{U}M$ is, via exp, in filtered bijection to the CGp $\mathcal{G}\widehat{U}M$, see (3.85). Hence the set M has a natural CGp-structure. Denote by $\mathcal{H}(M)$ this CGp, so $\exp : \mathcal{H}(M) \to \mathcal{G}\widehat{U}M$ is a CGp-isomorphism. Notice that $\mathcal{H}(M)$ and M have the same underlying filtered sets. Further, the identity element of $\mathcal{H}(M)$ is $0 \in M$, and the inverse of x in $\mathcal{H}(M)$ is $-x \in M$. This follows from the equations $\exp(0) = 1$ and $\exp(x)\exp(-x) = 1$, cf. (3.42).

It is clear that the multiplication of $\mathcal{H}(M)$ is given by the Hausdorff formula: $(x, y) \mapsto \mathbf{h}(x, y) = x + y + \frac{1}{2}[x, y] + \ldots (x, y \in M)$. Here substitution of x, y in $\mathbf{h}(\mathbf{X}, \mathbf{Y})$ is defined as in (3.29).

The following *Lazard - Mal'cev correspondence* is due to Lazard [Laz]. The rational case is due to Mal'cev.

THEOREM (3.87)
The functor $\mathcal{H} : \langle \mathrm{CLA} \rangle \to \langle \mathrm{CGp} \rangle$ *is an isomorphism of categories.*

Proof. We proceed (as far as possible) as in (3.86) to obtain the functor $\mathcal{H}^{-1}$. Let S be a set, and $A = \widehat{\mathsf{K}\mathsf{F}}(S)$. We show that $i : \mathsf{F}(S) \to \mathcal{G}A$ is a CGp-completion. Recall the canonical CHA-isomorphism $A \cong \mathsf{K}\langle\langle \mathbf{X}_S \rangle\rangle$, see (3.45). We have canonical graded LA-isomorphisms $\mathrm{gr}(\mathcal{G}A) = \mathrm{gr}(\mathcal{P}A) = \mathrm{gr}(\widehat{\mathsf{L}}(\mathbf{X}_S)) = \widehat{\mathrm{gr}}(\mathsf{L}(\mathbf{X}_S))$. It is well-known that $\mathrm{gr}(\mathsf{F}(S)) = \mathsf{L}\mathsf{F}_{\mathrm{ab}}(S)$, and similarly $\widehat{\mathrm{gr}}(\mathsf{L}(\mathbf{X}_S)) = \widehat{\mathsf{L}}(\mathsf{F}_{\mathrm{ab}}(S) \otimes \mathsf{K})$. Under these isomorphisms $\mathrm{gr}^1(i)$ is identified with the obvious map $\mathsf{F}_{\mathrm{ab}}(S) \to \mathsf{F}_{\mathrm{ab}}(S) \otimes \mathsf{K}$. Hence i has the characteristic property of a CGp-completion.

Hence $\widehat{\mathsf{F}}(S) = \mathcal{G}A$ has a natural CLA-structure, and with this structure we denote it by $\mathcal{H}^{-1}(\widehat{\mathsf{F}}(S))$. (This proves the desired category isomorphism on the full sub-categories of *rigid free* objects).

Let now H be any CGp. Then we define new operations $(H, +, [\,])$ as follows. For $x, y \in H$ there is a unique CGp-map $\phi : \widehat{\mathsf{F}}(s, t) \to H$ mapping the generators to x, y. Using the CLA-structure on $\widehat{\mathsf{F}}(s, t)$ declared above, we put $x + y = \phi(s + t)$ and $[x, y] = \phi([s, t])$. The Lie algebra relations for $\mathcal{H}^{-1}(H) = (H, +, [\,])$ follow from the corresponding relations for the generators of (rigid) free complete groups, regarded as Lie algebras; where it suffices to consider the case of 2 and 3 generators. Similarly, using that $F^n\widehat{\mathsf{F}}(s, t)$ is the complete subgroup generated by $\Gamma^n\mathsf{F}(s, t)$ (see (3.88)), one shows that $x + y = (xy)u$ with $u \in F^{n+1}H$ if $y \in F^nH$, and $[x, y] = (x, y)v$ with $v \in F^{i+j+1}H$ if $x \in F^iH$, $y \in F^jH$. Hence $\mathcal{H}^{-1}(H)$ is a *filtered* Lie algebra. These relations also imply that $\mathcal{H}^{-1}(H) \to H$, $x \mapsto x$ induces a graded LA-map (over $\mathbb{Z}$)

$$\mathrm{gr}(\mathcal{H}^{-1}(H)) \to \mathrm{gr}(H)$$

which is clearly bijective. Thus $\mathrm{gr}^n(\mathcal{H}^{-1}(H))$ is a K_n-module. We also get a canonical bijection $H/F^mH \approx \mathcal{H}^{-1}(H)/F^m\mathcal{H}^{-1}(H)$. We conclude $\mathcal{H}^{-1}(H) = \lim\{\mathcal{H}^{-1}(H)/F^n\mathcal{H}^{-1}(H)\}$ as a set, hence as a module. This completes the proof that $\mathcal{H}(H)$ is a CLA. It remains to show that $\mathcal{H}$ and $\mathcal{H}^{-1}$ are inverse functors. This is clear by definition, we omit the details. □

Remarks. This theorem corresponds to [Laz|II, Thm. (4.2), (4.3)]. Also Lazard makes use of Magnus algebras (formal power series rings), but *not* of their natural diagonals.

The theorem implies that for *any* CGp H, the canonical map $H \to \mathcal{G}\widehat{\mathsf{K}}H$ is an isomorphism, see below. Conversely, the theorem follows easily from this, cf. (3.86).

Notice that $\widehat{\mathsf{F}}(S) \to \mathcal{G}\mathsf{K}\langle\langle \mathbf{X}_S \rangle\rangle$, $s \mapsto \exp(\mathbf{X}_s)$ is a CGp-isomorphism, see proof.

Explicit formulas (3.88)

Recall that $\widehat{\mathsf{L}}(\mathbf{X}, \mathbf{Y}) = \prod_{n=1}^{\infty} \mathsf{L}^n(\mathbf{X}, \mathbf{Y}) \otimes \mathsf{K}_n$. It is thus clear that to any distinguished basis of $\mathsf{L}^n(\mathbf{X}, \mathbf{Y})$ (for all n) there corresponds a distinguished form of the Hausdorff series. There are explicit formulas ([Se], [Bki]).

Consider $\widehat{\mathsf{F}}(s,t) = \mathcal{G}\mathsf{K}\langle\langle \mathbf{X}, \mathbf{Y} \rangle\rangle$ as the free complete group generated by $s = \exp(\mathbf{X}), t = \exp(\mathbf{Y})$. By definition, $s+t = \exp(\mathbf{X}+\mathbf{Y})$ and $[s,t] = \exp([\mathbf{X}, \mathbf{Y}])$. Any element of $\widehat{\mathsf{F}}(s,t)$ may be written as the limit of a generic sequence $\prod_{n=1}^{\infty} z_n^{\alpha_n}$ where $z_n \in \Gamma^n \mathsf{F}(s,t)$ and $\alpha_n \in \Pi_n^{-1}\mathbb{Z}$. (This is the analog of (3.12) for filtered groups, notice that $\mathsf{F}(s,t) \to \widehat{\mathsf{F}}(s,t)$ is injective and filtration inducing). Lazard gives a method to compute explicit formulas

$$s+t = \prod_{n=1}^{\infty} u_n^{\alpha_n} = s\,t\,(s,t)^{-\frac{1}{2}} \ldots \qquad [s,t] = \prod_{n=1}^{\infty} v_n^{\beta_n} = (s,t)\,((s,t),t)^{\frac{1}{2}} \ldots$$

We point out that such explicit formulas are not needed for our purposes.

Now we have all ingredients to form the following diagram of functors (each having a canonical adjoint).

(3.89)

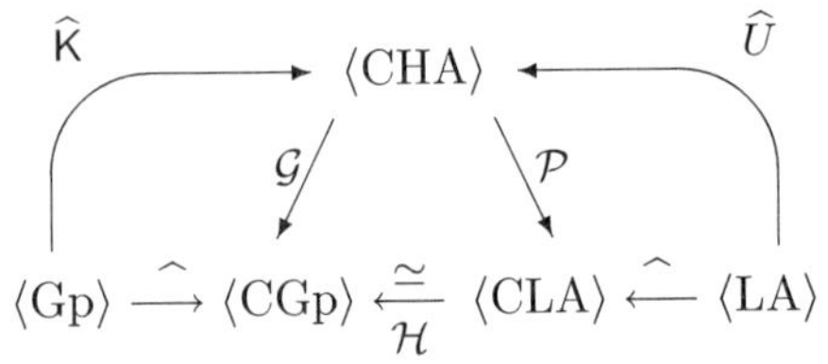
$\widehat{\mathsf{K}}$
⟨CHA⟩
$\widehat{U}$
$\mathcal{G}$
$\mathcal{P}$
⟨Gp⟩
⟨CGp⟩
$\cong$
$\mathcal{H}$
⟨CLA⟩
⟨LA⟩

THEOREM (3.90)

(a) *This diagram commutes up to canonical isomorphism:* $\widehat{G} \cong \mathcal{G}\widehat{\mathsf{K}}(G)$, $\widehat{L} \cong \mathcal{P}\widehat{U}(L)$, *and* $\mathcal{H}(\mathcal{P}A) \cong \mathcal{G}A$.

(b) $\mathcal{G}$ (resp. $\mathcal{P}$) *is left inverse to its left adjoint* $\widehat{\mathsf{K}}$ *(resp.* $\widehat{U}$*) up to canonical isomorphism:* $H \cong \mathcal{G}\widehat{\mathsf{K}}(H)$, *and* $M \cong \mathcal{P}\widehat{U}(M)$.

Note that in (b) the functors $\widehat{\mathsf{K}}$, $\widehat{U}$ are *not* that in the diagram, see (3.83).

Proof. It is clear that $\widehat{U} \cong \widehat{U}\,\widehat{}$, since the right adjoints of these functors coincide ($\mathcal{P} = \phi\mathcal{P}$ where ϕ forgets the filtration). We already know that $\mathrm{Id} \cong \mathcal{P}\widehat{U}$, which is the second assertion of (b). This implies the second assertion of (a): $\mathcal{P}\widehat{U} \cong \mathcal{P}(\widehat{U}\,\widehat{}\,) \cong \widehat{}\,$.

By definition the exponential map is a natural CGp-isomorphism

$$\exp : \mathcal{H}(\mathcal{P}A) \xrightarrow{\cong} \mathcal{G}A\,, \qquad A \in \langle\mathrm{CHA}\rangle\,.$$

Since the (left) adjoint of a functor is essentially unique, and $\mathcal{H}$ an isomorphism, it follows that $\widehat{\mathsf{K}}\mathcal{H} \cong \widehat{U}$. Moreover the second assertion of (b) implies the first: $\mathrm{Id} \cong \mathcal{G}\widehat{\mathsf{K}}$. As above the latter implies $\mathcal{G}\widehat{\mathsf{K}} \cong \widehat{}\,$, and the proof is complete. □

Remarks (3.91)

(a) We see that a CHA A is normal iff $A \cong \widehat{\mathsf{K}}H$ for some CGp H. Moreover, for any group G we have that $\widehat{\mathsf{K}}G = \widehat{\mathsf{K}}\widehat{G}$ (see proof) is normal. Hence, if we consider diagram (3.89) with $\langle \mathrm{CHA} \rangle$ = normal CHAs, then all functors are still defined, and the triangle in the middle consists of category equivalences. This implies easily that any normal CHA has a unique *antipodism* (i.e. it is a group object over the category of complete (cocommutative) coalgebras, where the product is $\widehat{\otimes}$). The category equivalence may be restricted to the full sub-categories of *rigid* objects, or to those objects whose associated graded module is *flat*. This follows from (3.85) and §2, N°4.

(b) Recall that over $\mathbb{Q}$ *all* CHAs are normal, so that the three categories in the middle of the diagram are equivalent. When restricted to the rigid objects, this equivalence is proved by Quillen, using different arguments. In particular, he does not construct the left adjoints of $\mathcal{G}$, $\mathcal{P}$ as we do. Instead he proves that $H \mapsto \lim\{\widehat{\mathsf{K}}\phi(H/F^{n+1}H)\}$ is quasi-inverse to $\mathcal{G}$, where ϕ forgets the filtration and the limit is to be taken in $\langle \mathrm{CHA} \rangle$. The above theorem implies that this functor coincides with our $\widehat{\mathsf{K}}$, but this does only work over $\mathbb{Q}$, and assuming rigidity.

COROLLARY (3.92)

Let S be a set. The canonical group map $\mathsf{F}(S) \to \mathcal{H}\widehat{\mathsf{L}}(S)$, $s \mapsto s$, induces a CGp-isomorphism $\widehat{\mathsf{F}}(S) \cong \mathcal{H}\,\widehat{\mathsf{L}}(S)$, whence the CLA-isomorphism $\mathcal{H}^{-1}\widehat{\mathsf{F}}(S) \cong \widehat{\mathsf{L}}(S)$.

Proof. Using (3.45) we obtain canonical CGp-isomorphisms

$$\widehat{\mathsf{F}}(S) \cong \mathcal{G}\,\widehat{\mathsf{K}}\,\mathsf{F}(S) \cong \mathcal{G}\,\widehat{U}\mathsf{L}(S) \cong \mathcal{H}\,\widehat{\mathsf{L}}(S)\,.$$

One readily checks that it is induced by $s \mapsto s$. □

Recall that the main result of this section states that for any CGp H the natural map $H \to \mathcal{G}\widehat{\mathsf{K}}H$ is a filtered isomorphism. Notice that this means the following: the map $H \to \widehat{\mathsf{K}}H$ is injective and induces the filtration of H, moreover every grouplike element comes from H.

We give some applications for group rings. Let $i : H \hookrightarrow \mathsf{K}H$ be the natural multiplicative map. Recall that $\mathsf{K}H$ has the associated filtration (3.84).

COROLLARY (3.93)

(a) *The natural map $H \hookrightarrow \mathsf{K}H$ is inducing: $F^nH = H \cap (1 + F^n\mathsf{K}H)$.*

(b) *If $F^{r+1}H = 1$ for some r, then $H \to \mathsf{K}H/F^{r+1}\mathsf{K}H$ is injective and inducing.*

Proof. Since $H \to \widehat{\mathsf{K}}H$ is inducing, the canonical map $\mathrm{gr}(H) \to \mathrm{gr}(\widehat{\mathsf{K}}H)$ is injective. It factors through $\mathrm{gr}(H) \to \mathrm{gr}(\mathsf{K}H)$, hence the latter is injective and the assertions follow. □

COROLLARY (3.94)

Let G be a filtered group such that $\mathrm{gr}\,G$ is torsion-free. Then $G \hookrightarrow \mathsf{K}G$ is inducing. □

In fact, by assumption $\mathrm{gr}(G) \to \mathrm{gr}(\widehat{G})$ is injective, hence (3.93) (a) yields the result. If the standard filtrations are considered, this yields a (well-known) solution, for certain groups, of the classical dimension subgroup problem.

As another application we prove (a special case of) the following representation theoretic result.

Theorem (3.95)
Let $\mathsf{K} = \Pi^{-1}\mathbb{Z}$. *Let* N *be a filtered group with* $F^{r+1}N = 1$, *and suppose* N *is* Π*-torsion-free. Then* $N \to \mathsf{K}N/F^{r+1}\mathsf{K}N = A$ *is injective.*
If N *is finitely generated, then* A *is finitely generated as* K*-module.*

Proof. *Let us assume in addition that* $\mathrm{M}_r \subset \Pi$ (where M_r = primes $\leqslant r$). Then the first assertion follows follows from (3.93) (b) and naturality, since by assumption $N \to \widehat{N}$ is injective (for a suitable choice of ring system we have here $\widehat{N} = N \otimes \mathsf{K}$). If N is finitely generated then so is $\mathrm{gr}(N)$, but the image of $\mathrm{gr}(N) \otimes \mathsf{K}$ generates $\mathrm{gr}(\mathsf{K}N)$, hence $\mathrm{gr}(A)$ and thus A is a finitely generated K-module. □

Hence a Π*-torsion-free, finitely generated, nilpotent group has a faithful unipotent representation over* $\Pi^{-1}\mathbb{Z}$ (on some finitely generated module). To see this, use the obvious filtration of A, and the embedding of A into $\mathrm{End}(A)$ via left multiplication.

For an elementary proof in the rational case (and for the standard filtrations), see [Q | p. 276 (3.6) (a)]. In fact, Quillen uses this result (and the Lie algebra analog — Ado's theorem) to prove a version of the Mal'cev correspondence, cf. (3.91) (b). One can prove the above theorem (in its stated generality) by using essentially Quillen's arguments. We do not know whether one can deduce from it the Lazard - Mal'cev correspondence. Under the obvious flatness assumptions however this is possible.

N° 4 The Quillen functor

Recall that K is a subring of $\mathbb{Q}$. By definition the *Quillen functor* λG carries any 2-reduced space X to the normalization of the simplicial (complete) Lie algebra $\mathcal{P}_{\cdot}\widehat{\mathsf{K}}_{\cdot}\, GX$ (cf. Chp. 4, § 3). By the commutativity of diagram (3.89) this sCLA is canonically isomorphic to $\mathcal{H}_{\cdot}^{-1}(\widehat{G}X)$.

The latter can be described in a more illuminating way, resembling the definition of Kan's loop group.

Theorem (3.96)
Let X *be a reduced space. Then* $\mathcal{L}X := \mathcal{H}_{\cdot}^{-1}(\widehat{G}X)$ *is the following (rigid) free sCLA:*

$$(\mathcal{L}X)_{n-1} = \widehat{\mathsf{L}}(X_n \smallsetminus s_0(X_{n-1})) = \widehat{\mathsf{L}}(X_n)//\widehat{\mathsf{L}}(s_0(X_{n-1}))$$

with s_i, d_j $(j \neq 0)$ *induced by* s_{i+1}, d_{j+1} *in* X, *and* $d_0(x) = \mathbf{h}(d_0x, -d_1x)$ *where* $\mathbf{h}$ *is the Hausdorff series.* □

The last term in the above equation denotes a cokernel in $\langle \mathrm{CLA} \rangle$.

The theorem is clear by the definition of GX and (3.92). Recall that $x^{-1} = -x$ in $\mathcal{H}\widehat{\mathsf{L}}(S)$. The Quillen functor is thus identified with $\lambda GX = N(\mathcal{L}X)$. The dg Lie algebra structure is obtained by "shuffling together" the simplicial Lie algebra structure of $\mathcal{L}X$.

Remarks (3.97)
(a) In the context of rational models (Chp. 4) we see that for computations in a finite range one may replace $\mathcal{L}X$ by a nilpotent quotient of a free sLA as follows. By Curtis' connectivity theorem (A.30) one may show that $F^{n+1}\mathfrak{m}$ is n-connected

for every reduced free sCLA $\mathfrak{m}$; cf. proof of (4.83). Thus from (the normalization of) the nilpotent sLA $\mathcal{L}X/F^{n+1}\mathcal{L}X$ we can build the n-th stage of a Quillen model for X. This is a quotient of a free sLA (by Γ^{n+1}); it is obtained by applying $\mathcal{H}^{-1}$ to the (rigid, discrete) sCGp $\widehat{G}X/F^{n+1}\widehat{G}X = (GX/\Gamma^{n+1}GX) \otimes \mathbb{Q}$ which is (necessarily) filtered by the lower central series.

(b) In case $\mathsf{K} = \mathbb{Z}$ the functor $\mathcal{H}^{-1}_{\cdot}(\widehat{G}X)$ is considered in [Dw] to construct his refinement of the Quillen model. By our results this functor is just a special case of the Quillen functor when it is defined over any ring.

Appendix C Filtered Modules

Here we study (complete) filtered modules and their tensor product in a more systematic way. We give proofs of certain statements in § 1, N° 2.

N° 1 Filtered vs. cofiltered modules

Recall that $\langle \mathrm{fil} \rangle$ denotes the category of (decreasing) filtered modules. We discuss its relation to cofiltered modules, where a cofiltered module is sort of a tower of modules. Then the notion of completeness is reflected in this context, where we consider the case of a constant ring system first.

Recall that a Hausdorff-complete module is a filtered module W such that the canonical module map $W \to \lim\{W/F^kW\}$ is an isomorphism. These are precisely the complete modules in the sense of (3.7), when the ring system is constant. Note that in this case the completion of a filtered module V is simply given by $\widehat{V} = \lim\{V/F^nV\}$ with $F^n\widehat{V} = \mathrm{Ker}(\widehat{V} \twoheadrightarrow V/F^nV)$. To see that $\widehat{V}$ is indeed Hausdorff-complete one only has to observe that the obvious map $c : V \to \widehat{V}$ induces isomorphisms $V/F^kV = \widehat{V}/F^k\widehat{V}$. The latter also yields an isomorphism $\mathrm{gr}\, V = \mathrm{gr}\, \widehat{V}$. Then it is easy to see that $c : V \to \widehat{V}$ has the desired universal property.

Since we consider the case of a constant ring system first, $\langle \mathrm{cpl} \rangle$ is the category of Hausdorff-complete modules.

Definition (C.1)

A *tower of modules* is a sequence $\{T_n, q_{n+1}\}$ of modules T_n and surjective module maps $q_{n+1} : T_{n+1} \twoheadrightarrow T_n$, where $n \geqslant 0$. With the evident notion of *tower map* these form a category $\langle \mathrm{cpl}' \rangle$. (This notation is motivated by (C.2)).

A *cofiltered module* is a module T, together with a tower of modules $\{T_n, q_{n+1}\}$ and surjective module maps $q'_n : T \twoheadrightarrow T_n$ such that $q_{n+1}q'_{n+1} = q'_n$ $(n \geqslant 0)$. With the evident notion of *cofiltered map* these form a category $\langle \mathrm{fil}' \rangle$. We identify $\langle \mathrm{cpl}' \rangle$ with the full subcategory of $\langle \mathrm{fil}' \rangle$ consisting of all cofiltered modules $T, \{T_n, q_{n+1}, q'_{n+1}\}$ such that the obvious map $q' : T \to \lim\{T_n, q_{n+1}\}$ is an isomorphism.

Theorem (C.2)

Associating to any filtered module V *the obvious cofiltered module* $V, \{V_n\}$ *where* $V_n = V/F^{n+1}V$, *settles an equivalence of categories* Tow : $\langle \mathrm{fil} \rangle \approx \langle \mathrm{fil}' \rangle$. — *It restricts to an equivalence* $\langle \mathrm{cpl} \rangle \approx \langle \mathrm{cpl}' \rangle$.

Proof. The proof is straightforward. A weak inverse of Tow is the functor Inv which carries the cofiltered module $T, \{T_n\}$ to the filtered module T with $F^nT = \operatorname{Ker}(T \twoheadrightarrow T_n)$. □

Definition (C.3)

Let a ring system K_* be given. Then $\langle \mathrm{cpl} \rangle$ denotes the category of complete modules, in the sense of (3.7); and $\langle \mathrm{cpl}' \rangle$ denotes the full subcategory of $\langle \mathrm{fil}' \rangle$ consisting of all towers of modules $\{T_n, q_{n+1}\}$ such that $\operatorname{Ker}(T_n \twoheadrightarrow T_{n-1})$ is a K_n-module.

Let $\langle \mathrm{cpl}'' \rangle$ be the following category. An object (called a *system*) is a sequence $\{X_n, \rho_{n+1}\}$ where X_n is a K_n-module, and $\rho_{n+1} : X_{n+1} \twoheadrightarrow X_n \otimes \mathsf{K}_{n+1}$ a surjective module map ($n \geqslant 0$). The morphisms are evident.

THEOREM (C.4)

There are canonical equivalences of categories

$$\langle \mathrm{cpl} \rangle \rightleftarrows \langle \mathrm{cpl}' \rangle \rightleftarrows \langle \mathrm{cpl}'' \rangle .$$

Under this equivalence the completion functor $\widehat{\ }$: $\langle \mathrm{fil} \rangle \to \langle \mathrm{cpl} \rangle$ *corresponds, up to canonical isomorphism, to the functor which carries a filtered module V to the system* $\{V_n \otimes \mathsf{K}_n, q_{n+1} \otimes 1\} \in \langle \mathrm{cpl}'' \rangle$.

Proof. It is clear that the category of complete modules $\langle \mathrm{cpl} \rangle$ corresponds, under the equivalence (C.2), to $\langle \mathrm{cpl}' \rangle$. Let $\Phi : \langle \mathrm{cpl}' \rangle \rightleftarrows \langle \mathrm{cpl}'' \rangle : \Psi$ be defined as follows. Φ sends the tower $\{T_n, q_{n+1}\}$ to the system $\{T_n \otimes \mathsf{K}_n, q_{n+1} \otimes 1\}$; and Ψ sends $\{X_n, \rho_{n+1}\}$ to the tower $\{X'_n, \rho'_{n+1}\}$ where $X'_0 = X_0$ and X'_{n+1}, ρ'_{n+1} is defined inductively by the pull-back diagram (of modules)

$$\begin{array}{ccc} X'_{n+1} & \xrightarrow{\alpha_{n+1}} & X_{n+1} \\ {\scriptstyle \rho'_{n+1}} \big\downarrow & \text{pull} & \big\downarrow {\scriptstyle \rho_{n+1}} \\ X'_n & \xrightarrow[\eta\,\alpha_n]{} & X_n \otimes \mathsf{K}_{n+1} \end{array}$$

where $\eta : X_n \to X_n \otimes \mathsf{K}_{n+1}$ is the canonical map, and $\alpha_0 = \mathrm{id}$. Now $\Psi\Phi \cong \mathrm{Id}$ follows from the fact that a "reasonable" diagram of modules (i.e. a square diagram whose vertical maps are surjective) is a pull-back *iff* the induced map on the kernels of the vertical maps is an isomorphism. And $\Phi\Psi \cong \mathrm{Id}$ follows from the induced fact that $- \otimes \mathsf{K}_{n+1}$ preserves reasonable pull-backs. We leave the details to the reader.

For the second assertion it suffices to show that the functor $\langle \mathrm{fil}' \rangle \to \langle \mathrm{cpl}'' \rangle$, $V \mapsto \{V_n \otimes \mathsf{K}_n\}$, is left adjoint - left inverse to Ψ (followed by the embedding). This is straightforward. □

Remark (C.5)

The composite equivalence Inv $\Psi : \langle \mathrm{cpl}'' \rangle \to \langle \mathrm{cpl} \rangle$ relates the system $\{X_n, \rho_{n+1}\}$ to the complete module consisting of all coherent sequences in the given system, where a *coherent sequence* is a family $\{x_n\}$ with $x_n \in X_n$ and $\rho_{n+1}(x_{n+1}) = x_n \otimes 1$. Then the second assertion of the theorem says that for any filtered module V, we may identify $\widehat{V}$ with the set of all coherent sequences in the system $\{V_n \otimes \mathsf{K}_n, q_{n+1} \otimes 1\}$. One may verify this directly.

Remark (C.6)

We mention the above theorem for two reasons:

(i) It provides a description of the completion functor that may be applied when working with filtered objects over any abelian category (see N° 4), or with filtered groups (see § 3, N° 2).

(ii) The authors of [FU 2], [Shu] work with the category $\langle \text{cpl}'' \rangle$, but do not even mention the equivalent $\langle \text{cpl} \rangle$. For a discussion to what extend this equivalence is compatible with monoidal structure, see (C.25).

N° 2 Normal maps and exactness

The category of filtered modules is additive; it has (co)kernels and (co)products — hence arbitrary (co)limits. However it is not abelian: monomorphisms (injections) need not be kernels, and epimorphisms (surjections) need not be cokernels. Similarly with complete modules. Hence one is led to the notion of 'normal' map; the terminology is borrowed from group theory where the normal subgroups are precisely the kernels. Let us begin however with ...

Coproducts and products (C.7)

Denote by $\bigoplus_i V_i$ (resp. $\prod_i V_i$) the obvious direct sum (resp. product) of a given family V_i $(i \in I)$ in $\langle \mathtt{fil} \rangle$. When the V_i are complete, $\widehat{\bigoplus_i} V_i := (\bigoplus_i V_i)\widehat{\ }$ is their coproduct in $\langle \text{cpl} \rangle$. It is clear that the completion functor preserves coproducts.

Now suppose that $\forall_{k \geqslant 0}\, V_i = F^k V_i$ except for a finite number of indices i. Then there is a canonical filtered isomorphism $\widehat{\bigoplus_i} V_i = \prod_i V_i$. The details are omitted.

(Co-)kernels (C.8)

It is clear that $\langle \mathtt{fil} \rangle$ is an additive category. Let $f : V \to W \in \langle \mathtt{fil} \rangle$. Then $\ker(f)$ exists: it is the usual kernel $f^{-1}(0)$, endowed with the induced filtration. Also $\operatorname{coker}(f)$ exists: it is the usual cokernel $f/\operatorname{Im}(f)$, endowed with the produced filtration.

The kernel of a map in $\langle \text{cpl} \rangle$ is defined as in $\langle \mathtt{fil} \rangle$. The cokernel of a map in $\langle \text{cpl} \rangle$ is defined as follows: take the cokernel in $\langle \mathtt{fil} \rangle$ and then apply completion. (One can also first complete the image (3.11), then factor it out and put the produced filtration on the quotient).

Special filtered maps (C.9)

One checks the following characterizations of notions introduced in (3.6). One may also read this as definitions. For a categorical interpretation see (C.17). A filtered map $f : V \to W$ is ...

normal $\Leftrightarrow$ $f : F^n V \to F^n W \cap \operatorname{Im}(f)$ is surjective for all n;

normal injective $\Leftrightarrow$ normal and injective $\Leftrightarrow$ $F^n V = f^{-1}(F^n W)$;

normal surjective $\Leftrightarrow$ normal and surjective $\Leftrightarrow$ surjective in each filtration;

normal bijective $\Leftrightarrow$ *filtered isomorphism* $\Leftrightarrow$ normal and bijective $\Leftrightarrow$ bijective in each filtration.

Examples (C.10)

(a) If V, W are splittable (3.15), and f is compatible with suitable splittings, then f is normal.

(b) If f is inducing or producing (3.6), then f is normal.

(c) If there is a filtered map $\pi : W \to V$ such that $f\pi f = f$, then f is normal.

(d) The class of normal injections (resp. surjections) is closed under composition (but not so with normal maps).

THEOREM (C.11)

Let $f : V \to W$, $g : W \to U \in \langle \mathrm{fil} \rangle$.

(a) *If f is normal surjective then* $\operatorname{gr} f$ *is surjective (and conversely provided W is separated, V Hausdorff-complete).*

(b) *If g is normal injective then* $\operatorname{gr} g$ *is injective (and conversely provided W is separated).*

Remark (C.12)

One may prove directly that g is inducing (3.6) *iff* $\operatorname{gr} g$ is injective. This implies (b) since g inducing and W separated implies g normal injective. A proof of the theorem is given below.

Definitions (C.13)

The sequence (f, g) is *normal exact* if $\operatorname{Im}(f) = \operatorname{Ker}(g)$ (as submodules of W) and both f, g are normal. — The sequence (f, g) is *short (normal) exact* if $\operatorname{Im}(f) = \operatorname{Ker}(g)$ and f is normal injective and g is normal surjective.

By the following lemma, (f, g) is short exact iff it is *short exact in each filtration*, i.e. $(F^n f, F^n g)$ is short exact for all $n \geqslant 0$.

LEMMA (C.14)

The sequence (f, g) is exact in each filtration iff $\operatorname{Im}(f) = \operatorname{Ker}(g)$ *and f is normal.*

THEOREM (C.15)

If (f, g) is normal exact then $(\operatorname{gr} f, \operatorname{gr} g)$ *is exact (and conversely provided W is separated, V Hausdorff-complete, and $gf = 0$).*

Proofs. The lemma is easily checked, and (C.11) (a)–(b) are special cases of (C.15). For (C.15) ($\Leftarrow$) we use (3.10). — Let $w = w_n \in F^n W$, $g(w_n) = 0$. Then $(\operatorname{gr}^n g)[w_n] = 0$ gives $v_n \in F^n V$, $w_{n+1} \in F^{n+1} W$ with $f(v_n) = w_n - w_{n+1}$. Now $g(w_{n+1}) = 0$ and we proceed by induction. We get a generic sequence $\sum_{i=n}^{\infty} v_i$ whose image under f is $\sum w_i - w_{i+1}$ which clearly has w as limit — uniquely as W is separated. As V is Hausdorff-complete, the sequence in V has a limit v (necessarily in filtration n) and $f(v) = w$. — Hence (f, g) is exact in each filtration, and f is normal by (C.14). In order to show g normal, one only has to observe that normality means the following:

$$w \in F^n W,\ g(w) \in F^{n+1} U \Rightarrow \text{there is } w' \in F^{n+1} W \text{ with } g(w) = g(w').$$

Using this, also (C.15) ($\Rightarrow$) presents no difficulty. □

COROLLARY (C.16)

(a) *If $f \in \langle \mathrm{fil} \rangle$ is normal surjective (normal injective, normal), so is $\widehat{f}$.*

(b) *If (f, g) is normal (short) exact, so is $(\widehat{f}, \widehat{g})$.* □

Categorical interpretation (C.17)

We give a categorical interpretation of the various notions discussed above. Suppose we are given an additive category with kernels and cokernels. Then any morphism $f : V \to W$ may be factorized into $f = i \circ f' \circ q$, where $q = \operatorname{coim} f := \operatorname{coker}(\ker f)$ and $i = \operatorname{im} f := \ker(\operatorname{coker} f)$. Here f' is uniquely determined, and different choices of (co)kernels yield canonically isomorphic factorizations. We define: f is *normal* if f' is an isomorphism. — f is a *kernel* if f is the kernel of some morphism $W \to U$, or equivalently, if $f = \ker(\operatorname{coker} f)$, or equivalently, if f is a normal monomorphism. — The dual notion is that of a *cokernel*. Note that f is normal iff it has a factorization into a cokernel followed by a kernel.

If $f = hg$ is a kernel and h a monomorphism (i.e. the kernel is zero), then g is a kernel. A morphism which admits a retraction is a direct summand, and thus a kernel. A kernel which is also an epimorphism, is an isomorphism. These assertions are easily verified, and so is the following:

In $\langle \mathrm{fil} \rangle$, *or* $\langle \mathrm{cpl} \rangle$, *the normal morphisms, kernels, cokernels, isomorphisms are,* respectively, *the normal maps, normal injections, normal surjections, normal bijections, as given by* (C.9). — Moreover, *composable morphisms* (f, g) *satisfy* $f = \ker(g)$ *and* $g = \operatorname{coker}(f)$ *iff* (f, g) *is short exact in the sense of* (C.13).

N° 3 Filtered tensor product

Recall the definition of the filtered tensor product (of filtered modules), see § 1, N° 2.

In some of the results on filtered tensor products one encounters the condition that a filtered module V be such that $\operatorname{gr} V$ is flat. By standard arguments one shows that this is equivalent to V/F^nV being flat for all n. Hence $\operatorname{gr} V$ flat implies V flat, if one of the following holds:

(1) the ground ring is hereditary and V separated;

(2) V is discrete (i.e. the filtration is finite).

Definition (C.18)

Let V be a filtered module. Define the filtered module $\mathbf{F}^kV$ to be the submodule $F^kV \subset V$ endowed with the induced filtration. Define the filtered module V_k to be the module $V/F^{k+1}V$ endowed with the produced filtration. (It is easy to describe these filtrations explicitly). Then we have, by (C.14), for each k the short exact sequence

$$\mathbf{F}^{k+1}V \hookrightarrow V \twoheadrightarrow V_k \qquad \text{in } \langle \mathrm{fil} \rangle .$$

We also have the short exact sequence

$$\mathbf{F}^{k+1}V \hookrightarrow \mathbf{F}^kV \twoheadrightarrow \mathbf{gr}^kV \qquad \text{in } \langle \mathrm{fil} \rangle$$

where $\mathbf{gr}^kV$ is the module gr^kV endowed with the produced filtration ($F^i = F^0$ for $i \leqslant k$ and $F^i = 0$ else).

THEOREM (C.19)

Let V, X *be filtered modules. Then the obvious map* $\operatorname{gr}(\rho_m \otimes X) : \operatorname{gr}(V \otimes X) \twoheadrightarrow \operatorname{gr}(V_m \otimes X)$ *is bijective in degrees* $\leqslant m$ (resp. $\leqslant m+1$ *if* $\operatorname{gr}^0 X = 0$).

Proof. This map is surjective by (3.17), or (C.21) (a) below. The sequence

$$(*) \qquad \mathbf{F}^{m+1}V \otimes X \to V \otimes X \twoheadrightarrow V_m \otimes X \qquad \text{in } \langle \mathrm{fil} \rangle$$

is an exact sequence of modules. It is thus exact in filtrations $\leqslant m+1$, since we have $F^0(\mathbf{F}^{m+1}V \otimes X) = F^{m+1}(\mathbf{F}^{m+1}V \otimes X)$. The latter follows, by surjectivity of

$$\mathrm{gr}(\mathbf{F}^{m+1}V) \otimes \mathrm{gr}(X) \to \mathrm{gr}(\mathbf{F}^{m+1}V \otimes X)$$

from the obvious relation $F^0(\mathbf{F}^{m+1}V) = F^{m+1}(\mathbf{F}^{m+1}V)$. Hence 'gr' applied to $(*)$ is exact in degrees $\leqslant m$, where $\mathrm{gr}(\mathbf{F}^{m+1}V \otimes X)$ is zero. This yields the first assertion, and the second follows by the same arguments. □

COROLLARY (C.20)

Let $f : V \to W$ be a filtered map, X a filtered module. If $\mathrm{gr}(f)$ is bijective in degrees $\leqslant m$, so is $\mathrm{gr}(f \otimes X)$.

If in addition $\mathrm{gr}^0 X = 0$, then $\mathrm{gr}(f \otimes X)$ is bijective in degrees $\leqslant m+1$.

Proof. By assumption $f_m : V_m \to W_m$ is a filtered isomorphism. Hence $f_m \otimes X : V_m \otimes X \to W_m \otimes X$ is a filtered isomorphism. The preceding result (applied to V, X resp. W, X) yields the assertion. □

PROPOSITION (C.21)

Let $f : V \to W$, $g : W \to U$ and $X \in \langle \mathrm{fil} \rangle$.

(a) *If g is normal surjective, so is $g \otimes X : W \otimes X \to U \otimes X$.*

(b) *Let $\mathrm{gr}\, X$ be flat. If f is normal, so is $f \otimes X$.*

(c) *Let $\mathrm{gr}\, X$ and X be flat. If f is normal injective, so is $f \otimes X$.*

(d) *Let $\mathrm{gr}\, X$ and X be flat. If (f, g) is short exact* (resp. *normal exact, exact in each filtration*)*, the same holds for $(f \otimes X, g \otimes X)$.*

Proof. Part (a) follows directly from the definition of the filtered tensor product. *Ad* (b): recall that f normal means $f = j \circ q$ with q normal surjective, j normal injective. Using (a) and the fact that the class of normal surjections is closed under composition, it suffices to show that $j \otimes X$ is normal. Now $\mathrm{gr}\, j$ is injective by (C.11) (b), hence so is $\mathrm{gr}(j) \otimes \mathrm{gr}(X) = \mathrm{gr}(j \otimes X)$, where the latter uses (C.24). This implies that $j \otimes X$ is inducing, hence normal, see (C.12). Thus (b) is proved, and (c) follows. (d) follows from (b) and (C.14). □

We use the preceding result to show that $F^k(V \otimes W)$ has a *natural* finite filtration, with smallest term $V \otimes F^k W$ and successive quotients $F^{k-j}V \otimes \mathrm{gr}^j W$ (up to canonical isomorphism), where $j = k-1, \ldots, 0$. This is an important means in the study of filtered simplicial (or dg) modules. We impose certain extra conditions here, cf. the introductory notes to the present subsection.

THEOREM (C.22)

Let V, W be filtered modules. Suppose that $\mathrm{gr}\, V$ and V are flat. For any $k \geqslant 1$, there are natural short exact sequences (of modules)

$$\begin{gathered} V \otimes F^k W \hookrightarrow X_1(V, W; k) \twoheadrightarrow F^1 V \otimes \mathrm{gr}^{k-1} W \\ X^{k-1}(V, W) \hookrightarrow X_2(V, W; k) \twoheadrightarrow F^2 V \otimes \mathrm{gr}^{k-2} W \\ \vdots \\ X_{k-1}(V, W; k) \hookrightarrow F^k(V \otimes W) \twoheadrightarrow F^k V \otimes \mathrm{gr}^0 W\,. \end{gathered}$$

Proof. By (C.21) (d) and (C.18) we have short exact sequences of filtered modules

$$V \otimes \mathbf{F}^{j+1}W \hookrightarrow V \otimes \mathbf{F}^{j}W \twoheadrightarrow V \otimes \mathbf{gr}^{j}W$$

where $j = k-1, \dots, 0$. Their restriction to filtration k yield the desired short exact sequences. In fact, by definition $F^k(V \otimes \mathbf{gr}^j W)$ is the image of the map $F^{k-j}V \otimes \mathrm{gr}^j W \to V \otimes \mathrm{gr}^j W$, but this map is injective since $V/F^{k-j}V$ is flat. □

Remark (C.23)

If in addition W is splittable (3.15), the short exact sequences in the above theorem are split. Hence there is a (non-natural) isomorphism

$$F^k(V \otimes W) \approx V \otimes F^k W \oplus \bigoplus_{j=0}^{k-1} F^{k-j}V \otimes \mathrm{gr}^j W\,.$$

This may be shown directly (at least under certain extra assumptions) using the decomposition of *filtered* modules: $V \otimes W = V \otimes \mathbf{F}^k W \oplus \bigoplus_{j=0}^{k-1} V \otimes \mathbf{gr}^j W$ which holds if W is splittable (recall that the filtered tensor product is additive).

Theorem (C.24)

The canonical surjective map $\pi : \mathrm{gr}(V)\otimes\mathrm{gr}(W) \to \mathrm{gr}(V\otimes W)$ *is an isomorphism if* $\mathrm{gr}\,V$ *is flat.*

Of course, the condition: $\mathrm{gr}\,V$ *flat*, can be replaced by: V, W *splittable*. There are very simple counterexamples showing that some extra condition is necessary.

Proof. Consider the "inclusion" diagram of modules given by the $A^{i,j} = F^iV \otimes F^jW$ $(i, j \geqslant 0)$, the "horizontal" maps $A^{i-1,j} \leftarrow A^{i,j} : h$ and the "vertical" maps $v : A^{i,j} \to A^{i,j-1}$ (induced by inclusion). The flatness assumption implies without difficulty that the horizontal maps are injective, and that all the square subdiagrams of the inclusion diagram are pull-backs. This means that the sequence

$$0 \to A^{i,j} \xrightarrow{v-h} A^{i,j-1} \oplus A^{i-1,j} \xrightarrow{(h,v)} A^{i-1,j-1}$$

is exact. An induction then shows that also

$$(*) \qquad 0 \to \bigoplus_{i=0}^{k-1} A^{i+1,k-i} \xrightarrow{\delta} \bigoplus_{i=0}^{k} A^{i,k-i} \xrightarrow{(\mathrm{incl})} A^{0,0} = V \otimes W$$

is exact where $\delta = v - h$ on each summand. We clearly have the exact sequence

$$(**) \qquad \bigoplus_{i=0}^{k} A^{i,k-i+1} \oplus A^{i+1,k-i} \xrightarrow{\bigoplus(v,h)} \bigoplus_{i=0}^{k} A^{i,k-i} \twoheadrightarrow \bigoplus_{i=0}^{k} \mathrm{gr}^i V \otimes \mathrm{gr}^{k-i} W$$

(a direct sum of $k+1$ exact sequences), which throws obvious surjective maps to the short exact sequence

$$F^{k+1}(V \otimes W) \hookrightarrow F^k(V \otimes W) \twoheadrightarrow \mathrm{gr}^k(V \otimes W)\,.$$

Since $(*)$ is exact, δ is the kernel of the middle surjection. One checks that δ factors through the first map in $(**)$. By a simple diagram chase this implies that the map in question (the right surjection) is injective. □

Filtered vs. cofiltered tensor product (C.25)

Can one give an interpretation of the filtered tensor product, in terms of cofiltered modules?—As in the above proof (consider the "inclusion" diagram), one proves that the submodules $F^{n+1}(V \otimes W) \subset \bigcap_{i=0}^{n}(F^{i+1}V \otimes W + V \otimes F^{n+1-i}W)$ of $V \otimes W$ coincide if $\operatorname{gr} V$ is flat. In other words, $F^{n+1}(V \otimes W)$ is the kernel of the obvious map $V \otimes W \to \bigoplus_{i=0}^{n} V_i \otimes W_{n-i}$. The image of this map is (always) the limit of the obvious diagram formed by the $V_i \otimes W_j$ $(i + j \leqslant n)$. We conclude that

$$(V \otimes W)_n = \lim_{i+j \leqslant n} V_i \otimes W_j \,.$$

The latter is the natural definition of a *cofiltered tensor product* (of cofiltered modules), which gives an additive monoidal structure on $\langle \mathrm{fil}' \rangle$. What is shown above means that the equivalence $\mathrm{Tow} : \langle \mathrm{fil} \rangle \approx \langle \mathrm{fil}' \rangle : \mathrm{Inv}$ is (additive) monoidal on certain full subcategories. More precisely, the canonical surjective cofiltered map

$$\phi : \mathrm{Tow}(V \otimes W) \longrightarrow \mathrm{Tow}(V) \otimes \mathrm{Tow}(W)$$

is an isomorphism if $\operatorname{gr} V$ is flat. Consequently, certain categories of filtered algebras are equivalent to certain categories of cofiltered algebras.

Now let a ring system be given. We obtain immediately an analogous result concerning the equivalences $\langle \mathrm{cpl} \rangle \approx \langle \mathrm{cpl}' \rangle \approx \langle \mathrm{cpl}'' \rangle$. In fact, the second equivalence in this sequence is monoidal (without any restriction) if we define a complete tensor product on $\langle \mathrm{cpl}'' \rangle$ by

$$(X \mathbin{\widehat{\otimes}} Y)_n = \lim_{i+j \leqslant n} X_i \otimes Y_j \otimes \mathsf{K}_n \,.$$

This definition is used in the papers cited in (C.6). We leave the details to the reader.

Remark (C.26)

We finally mention that for filtered modules V, W there is a canonical module(!) isomorphism $\lim\{V \otimes W / F^n(V \otimes W)\} = \lim\{V/F^nV \otimes W/F^nW\}$. In fact, $F^{2k}(V \otimes W) \subset \mathrm{Im}(F^kV \otimes W \oplus V \otimes F^kW \to V \otimes W) \subset F^k(V \otimes W)$; passing to quotients yields the assertion.

N° 4 Complete Differential Algebra

It is essentially trivial to generalize the concepts and results of this chapter (*except* those concerned with groups!) to the differential graded setting. One only has to read 'module' as 'chain complex'. In other words, instead of starting with the abelian monoidal category $\mathcal{M}od_{\mathsf{K}}$, we start with $\langle \mathrm{DG} \rangle$. Then $\langle \mathrm{fil} \rangle$ is the category of (descending) filtered chain complexes.

Here the tensor product of DG modules is as usual; the interchange map is defined using the sign rule, which is of importance in the definition of some species of DG algebras (e.g. commutative algebra, Lie algebra).

Filtered chain complexes (C.27)

Notice that a *filtered chain complex* (= filtered object over the category of chain complexes) is the same as a *DG filtered module* (= DG object over the category of filtered modules). Formally this is a category isomorphism compatible with tensor product. In fact, if V, W are filtered chain complexes (subscripts denote the degree),

$$\begin{aligned}
[F^k(V\otimes W)]_n &= [\operatorname{Im}(\textstyle\bigoplus_{i+j=k} F^iV\otimes F^jW \to V\otimes W)]_n \\
&= \textstyle\bigoplus_{\ell+m=n}\operatorname{Im}(\bigoplus_{i+j=k} F^iV_\ell\otimes F^jW_m \to V_\ell\otimes W_m) \\
&= \textstyle\bigoplus_{\ell+m=n} F^k(V_\ell\otimes W_m) \\
&= F^k(\textstyle\bigoplus_{\ell+m=n} V_\ell\otimes W_m) \\
&= F^k([V\otimes W]_n)
\end{aligned}$$

where in the last equation we regard V, W as DG filtered modules.

Of course the completion of a filtered chain complex V is given by $(\widehat{V})_n = (V_n)\widehat{\ }$, i.e. by degreewise application of the corresponding construction with filtered modules. Similarly, a filtered chain map is normal (injective, surjective) iff it is so in each degree, *etc.*

We give some more details, indicating further generalizations (by way of axiomatization) of the theory presented in this chapter.

A generalization (C.28)

Begin with any abelian category $\mathcal{M}$ whose objects play the role of *modules*. Let it be endowed with a (right exact) *tensor product* $\otimes$. Denote the unit object by $K\in\mathcal{M}$. Then $\mathcal{M}$ with the structure $(\otimes, K, \dots)$ is an *abelian monoidal category*; cf. Chp. 2 and [Mc3].

The category $\langle \mathrm{fil}\rangle$ and its (strongly) additive monoidal structure is defined as in § 1, N° 2. The above results of App. C resp. of § 1, N° 2 hold in the general context, although in some cases different arguments must be found. For instance, one has to abandon the use of topological notions (or introduce these in a different way). Given a set of projective generators (containing K) in $\mathcal{M}$, *free objects* are defined in $\mathcal{M}$, and then in $\langle \mathrm{fil}\rangle$. But actually this notion does not play a substantial role.

Now $X\in\mathcal{M}$ is Π-*uniquely-divisible* if multiplication with any Π-integer is an automorphism of X. We assume: (1) for any set of primes Π the full subcategory $\Pi^{-1}\mathcal{M}\hookrightarrow\mathcal{M}$ of Π-uniquely-divisible "modules" is reflective. This means that the embedding has a left adjoint - left inverse, denoted by Π^{-1}. (2) Π^{-1} is exact, and the canonical map $\Pi^{-1}(X\otimes Y)\to X\otimes\Pi^{-1}Y$ is an isomorphism. — As in § 1, N° 1 this implies that Π^{-1} has all desirable properties. In particular, $\Pi^{-1}X = X\otimes\Pi^{-1}K$.

An increasing sequence of sets of primes Π_* determines the sequence of subcategories $\mathcal{M}_*$ where $\mathcal{M}_n = \Pi_n^{-1}\mathcal{M}$. The latter is already determined by K_* where $K_n = \Pi_n^{-1}K$ (the "ring system"). In fact, $\mathcal{M}_n$ consists of all X such that $X\to X\otimes K_n$ is an isomorphism (the "K_n-modules"). Notice that $\mathcal{M}_n$ is an abelian monoidal category with unit object K_n.

Let a "ring system" K_* be given. The category $\langle\mathrm{cpl}\rangle$ and its (strongly) additive monoidal structure is defined as in § 1, N° 2. Since $\mathcal{M}$ has limits, a completion functor exists (use the construction presented in App. C, N° 1). Now various species of complete algebras are defined. Assume now that K_* is *mild*, i.e. K_n is M_n-uniquely-divisible where M_n = primes $\leqslant n$. Then most of § 2 goes through without change.

There is one difficulty: working over $\langle \text{cpl} \rangle, \widehat{\otimes} \ldots$, free Lie algebras over any complete module may be constructed, but it is not obvious that the free Lie algebra is a retract of the tensor algebra; cf. (3.53).

However when $\mathcal{M}$ is the category of chain complexes over some ring, there is no difficulty here (cf. proof of the PBW–Theorem for DG Lie algebras in [Q]).

Further generalization (C.29)

An even more general approach (which I would prefer) starts with any additive monoidal category $\mathcal{V}$, playing the role of $\langle \text{cpl} \rangle$. Recall that this is the setting of Chp. 2. The principal assumption on $\mathcal{V}$ is the following:

(M) There is a sequence of reflective (full) subcategories $\mathcal{V} = \mathcal{V}_0 \supset \mathcal{V}_1 \supset \ldots$ such that (1) $\mathcal{V}_n$ consists of M_n-uniquely-divisible objects (where M_n = primes $\leqslant n$), and (2) the tensor product on $\mathcal{V}$ restricts to $\otimes : \mathcal{V}_i \times \mathcal{V}_j \to \mathcal{V}_{i+j}$.

For example, if a *mild* ring system is given, $\langle \text{cpl} \rangle$ satisfies (M): here $\mathcal{V}_n$ is the category of complete modules V satisfying $V = F^n V$. In fact, for the definition of exponential-like maps we need that $W \in \mathcal{V}_1$ implies that $W^{\otimes n} \in \mathcal{V}_n$ is M_n-uniquely-divisible.

Again a lot of § 2 goes through without change, but it is likely that further assumptions are to be imposed on $\mathcal{V}$.

Increasing filtrations (C.30)

An *increasing filtered module* is a module V together with a sequence of submodules $F_0 V \subset F_1 V \subset \ldots \subset V$. It is said to be *exhaustive* if $V = \operatorname{colim} F^n V$. The category $\langle \mathtt{fil}^+ \rangle$ of increasing filtered modules (over some abelian category $\mathcal{M}$) is dual to the category of decreasing filtered modules (over $\mathcal{M}^{\text{op}}$), by (C.2). The results of N° 2 can thus be dualized. (In [Bki 2] some of these are proved for filtrations which are infinite in both directions).

Let us call V (as above) *complete* if $F_n V$ is a K_n-module. There is a completion functor $\langle \mathtt{fil}^+ \rangle \to \langle \text{cpl}^+ \rangle$, left adjoint - left inverse to the embedding. (One can work with exhaustive or non-exhaustive filtered modules).

Define a tensor product on $\langle \mathtt{fil}^+ \rangle$ as in § 1, N° 2. Then $\langle \text{cpl}^+ \rangle$ is endowed with the obvious complete tensor product. If the ring system is mild, this additive monoidal category has the property (M) mentioned in (C.29).

It is thus likely that the results of this chapter hold analogously if we work over $\langle \text{cpl}^+ \rangle$. Unfortunately, the mentioned dualism is *not* of great use here; in particular it is not compatible with monoidal structure in general, cf. (C.25).

Increasing complete (commutative) dg* algebras are used as models in "tame homotopy theory via polynomial forms", see [Maj], [Ste], [FU].

Chapter 4

Three Models for Spaces

In this chapter K is a field of characteristic 0, or at least a hereditary $\mathbb{Q}$-algebra (commutative, as always), where *hereditary* means that a submodule of a free module is free. Further generalizations, if possible, are indicated. Modules, tensor products etc. are over K unless otherwise specified. We use $H_*(X) = H_*(X;\mathsf{K})$ for the homology of the normalized chain complex $C(X) = C(X;\mathsf{K})$ regarded as a quotient of $\mathsf{K}X$. Analogously in the context of cohomology. Spaces are pointed and DG (co)algebras are (co)augmented, analogously for maps.

A map of spaces, or chain complexes, is said to be an *H_*-equivalence* if it induces isomorphisms via the functor H_*. Analogously for other functors.

§ 1 The Cellular Model

In this section K is any commutative ring. We use the notation introduced in the first section of Chp. 2.

N° 1 The homotopical category of dg algebras

Recall that $\langle \mathrm{dgA}\rangle_0$ is the category of augmented dg algebras over K. By definition, a map $f : B \to A \in \langle \mathrm{dgA}\rangle_0$ is *free* if there is a graded free module V and a graded algebra isomorphism $\alpha : A \approx B \amalg TV$ such that $\alpha f = \mathrm{incl}$. An object B is *free* if its unit is a free map.

For the notion of homotopy structure in a category, see the beginning of Chp. 1. Recall that a category with distinguished homotopy structure is called a homotopical category. Homotopical *sub*categories are defined in (1.7).

THEOREM (4.1)

The category $\langle \mathrm{dgA}\rangle_0$ has a canonical homotopy structure: the weak equivalences are the H_*-equivalences, cofibrant means free, and all objects are fibrant.

The homotopy relation is given by derivation homotopy.

Moreover $\langle \mathrm{dgA}\rangle_1 \subset \langle \mathrm{dgA}\rangle_0$ is a homotopical subcategory.

Thus we have the following explicit discription of homotopy in $\langle \mathrm{dgA}\rangle_0$ or $\langle \mathrm{dgA}\rangle_1$, cf. (2.8). Let B be free. Then $f \simeq g : B \rightrightarrows A$ iff there is a degree $+1$ map $H : B \to A$ satisfying

$$(4.2)\qquad \begin{aligned} dH(x) + H(dx) &= f(x) - g(x)\,, \\ H(xy) &= H(x)g(y) + (-1)^{|x|} f(x)H(y) \end{aligned} \qquad \text{for} \quad x, y \in B\,.$$

By the last assertion of the theorem, the derived category $\mathrm{Ho}\langle\mathrm{dgA}\rangle_1$ is just the full subcategory of $\mathrm{Ho}\langle\mathrm{dgA}\rangle_0$ having the same objects as $\langle\mathrm{dgA}\rangle_1$.

Proof. In fact $\langle\mathrm{dgA}\rangle_0$ is a model category where a cofibration is a free map, and a fibration a map which is surjective in degrees > 0. The verification of the axioms (A.9) is straightforward. Note that any free map is a transfinite composition of "single attachments" $B \hookrightarrow (B \amalg \mathsf{T}(x), d)$ where dx is any cycle of B, compare (4.21). Another point is that the (free) inclusion map $B \to B \amalg \mathsf{T}(V \oplus dV) = A$ is a weak equivalence where $dV = s_{-1}V$ is any (positive) graded free module. For this define an obvious derivation homotopy on A (zero on B), cf. [B].

There is (see below) a natural cylinder object IB for cofibrant B in $\langle\mathrm{dgA}\rangle_0$ with respect to which two maps $B \rightrightarrows A$ are homotopic iff they are derivation homotopic. (There is also a natural path object PA with that property [Mu].) This implies the second assertion.

It remains to show that $\langle\mathrm{dgA}\rangle_1$ is a homotopical subcategory. By (1.17) this is a consequence of the following three facts:

(i) Any $A \in \langle\mathrm{dgA}\rangle_1$ has a *reduced* free dgA model $LA \xrightarrow{\sim} A$.

(ii) There is an Eilenberg functor (1.14) $R_1 : \langle\mathrm{dgA}\rangle_0 \to \langle\mathrm{dgA}\rangle_1$.

(iii) If B is a *reduced* free dgA then also the cylinder object IB is reduced.

Indeed (i) is clear by construction, (ii) by (4.42), and (iii) by definition of IB. □

The cylinder object of Baues - Lemaire (4.3)

Let $B = (\mathsf{T}V, d)$ be a free dgA. Define $IB = (\mathsf{T}(V^0 \oplus V^1 \oplus sV),\, D)$ where V^0, V^1 are copies of V, and D is the unique derivation such that (i) the obvious inclusions $i_0,\ i_1 : B \rightrightarrows IB$ are dgA-maps, (ii) the unique (i_0, i_1)-derivation $S : B \to IB$ with $S(v) = sv$ $(v \in V)$ is a derivation homotopy $i_0 \simeq i_1$.

Define $q : IB \to B$ by $v^0,\ v^1 \mapsto v$ and $sv \mapsto 0$. There is a canonical derivation homotopy $h : i_1 q \simeq 1_{IB}$, hence q is an H_*-equivalence. Since clearly $(i_0, i_1) : B \amalg B \to IB$ is a free map, we see that (IB, i_0, i_1, q) is a cylinder object for B in the model category $\langle\mathrm{dgA}\rangle_0$, thus in the underlying homotopical category. Now composition with S settles the desired bijection

$$\{IB\text{-homotopies from } f \text{ to } g\} \approx \{\text{derivation homotopies from } f \text{ to } g\}\,.$$

For any free B the definition of IB depends upon the choice of a graded algebra isomorphism $\alpha_B : B \approx \mathsf{T}V$. But for some fixed choice (for each B) this cylinder object is natural. It follows that different choices yield isomorphic cylinder functors. For more details see [B], [An].

Remarks (4.4)

(a) In $\mathrm{Ho}\langle\mathrm{dgA}\rangle_0$ we have $[A, A'] = [\Omega BA, A'] = \mathrm{T}(BA, A')/_{\simeq}$ because $\Omega BA \xrightarrow{\sim} A$ is a cofibrant model, see (2.22) (a), (2.28). A twisting map $v \in \mathrm{T}(BA, A')$ is also called a *strongly homotopy multiplicative* map from A to A'.

(b) The arguments actually show that $\langle\mathrm{dgA}\rangle$ has canonical model structure inducing one in $\langle\mathrm{dgA}\rangle_0$. One may show that also $\langle\mathrm{dgA}\rangle_r$ $(r \geqslant 1)$ has canonical model structure.

(c) At least in case K is hereditary, also $\langle\mathrm{DGA}\rangle$ has canonical model structure. It "induces" the model structure in $\langle\mathrm{dgA}\rangle$ and also a model structure in $\langle\mathrm{dg}^*\mathrm{A}\rangle$. The second case (negative grading) is slightly more complicated; compare the proof in the commutative case (4.22).

N° 2 The homotopical category of dg Hopf algebras up to homotopy We apply the general considerations of Chp. 1, §3 to the homotopical category $\langle\text{dgA}\rangle_1$ which is also a monoidal category with the usual tensor product. The lemma below (*Künneth Theorem*) follows from the usual Künneth Formula in case the ground ring is hereditary (then complexes need not be bounded below). The general case follows from the spectral Künneth Formula [Mc 2].

Lemma (4.5)

If f, g are H_*-equivalences between bounded below chain complexes, then so is $f \otimes g$ — provided f or g is a map between degreewise flat objects. □

Hence $\langle\text{dgA}\rangle_1$ satisfies condition (T′) in Chp. 1, §3, N° 3. Here $\mathcal{E}$ is the full subcategory consisting of the degreewise flat objects. In fact $\mathcal{E}$ is clearly a left homotopical (thus localizing) subcategory containing the cofibrant objects and (thus) K, the tensor product restricts to $\mathcal{E}$ where it carries a pair of weak equivalences to a weak equivalence.

Hence $\text{Ho}\langle\text{dgA}\rangle_1$ has an (essentially well-defined) induced monoidal structure $\bar{\otimes}, \mathsf{K}, \ldots$ Note that $[f]\bar{\otimes}[g] = [f \otimes g]$ if f, g are maps of $\mathcal{E}$. Since in fact all dgAs of interest are degreewise flat we allow ourself to use the notation $\otimes$ also for the induced tensor product on $\text{Ho}\langle\text{dgA}\rangle_1$.

Using this structure there is a notion of *cocommutative comonoid up to homotopy* over $\langle\text{dgA}\rangle_1$. The following terminology is certainly more suggestive.

Definition (4.6)

Let $\langle\text{dgHh}\rangle_1 = \text{cComh}\mathcal{A}$ where $\mathcal{A} = \langle\text{dgA}\rangle_1$. An object of this category is called *(dg) Hopf algebra up to homotopy* (henceforth: *dgHh*). A dgHh is said to be *free*, etc. if it is so-called in $\mathcal{A}$. The free dgHhs form a full subcategory $\mathbf{dgHh}_1 \subset \langle\text{dgHh}\rangle_1$.

Note that dgHhs are assumed to be *reduced*.

In other words a dgHh is a pair (A, Ψ) where $A \in \mathcal{A}$ and $\Psi : A \to A \otimes A \in \text{Ho}\mathcal{A}$ is a homotopy Hopf diagonal — i.e. satisfies the relations for a cocommutative comonoid in $\text{Ho}\mathcal{A}$ (1.36). A dgHh-map $f : (A, \Psi) \to (B, \Psi')$ is a dgA-map satisfying $\Psi'[f] = ([f] \otimes [f])\Psi$.

Warning. In general, a map $\psi : B \to B \otimes B \in \mathcal{A}$ does *not* induce a homotopy diagonal on B. This is true however if B is flat since then $B \otimes B$ is also the tensor product in $\text{Ho}\mathcal{A}$. — Conversely, a homotopy diagonal on B need *not* be induced by such a map ψ. This is true however if B is cofibrant.

The following is an instance of (1.41).

Theorem (4.7)

The category $\langle\text{dgHh}\rangle_1$ has a canonical homotopy structure: the weak equivalences are the H_*-equivalences, cofibrant means free, and all objects are fibrant. The homotopy relation is given by derivation homotopy. Moreover $\text{Ho}\langle\text{dgHh}\rangle_1 = \text{cCom-Ho}\langle\text{dgA}\rangle_1$. □

Remark. For comparison with Anick's work ([An]) apply the discussion of individual cComhs (1.47) in the present situation. There is a slight simplification here since the unit object is terminal in $\mathcal{A}$. In fact the individual dgHhs are precisely the (free) 'Hopf algebras up to homotopy' used by Anick.

N° 3 The cobar - chain functor and the chain - loop functor

We show that for $X \in \mathcal{S}_2$ there is a natural H_*-equivalence from $\Omega C(X)$ to $C(GX)$.

We define the *cobar - chain functor* as the composite

$$\Omega C : \mathcal{S}_1 \longrightarrow \langle \mathrm{dgA} \rangle_0 . \tag{4.8}$$

Here $\mathcal{S}_1$ is the category of reduced spaces, $C : \mathcal{S}_1 \to \langle \mathrm{dgC} \rangle_1$ the normalized chain coalgebra functor and $\Omega : \langle \mathrm{dgC} \rangle_1 \to \langle \mathrm{dgA} \rangle_0$ the cobar construction (2.20). Notice that $\Omega C(X)$ is a free (cofibrant) dgA.

We define the *chain - loop functor* as the composite

$$CG : \mathcal{S}_1 \longrightarrow \langle \mathrm{dgA} \rangle_0 . \tag{4.9}$$

Here $G : \mathcal{S}_1 \to \langle \mathrm{sGp} \rangle$ is Kan's loop functor to the category of simplicial groups, and $C : \langle \mathrm{sGp} \rangle \to \langle \mathrm{dgA} \rangle_0$ is the normalized chain algebra functor.

For a more detailed discussion see App. B, N° 2. Let $t : C(X) \to \Omega C(X)$ be the universal twisting map and $\tau : X \to GX$ the universal twisting function [May].

THEOREM (4.10)

(a) *For $X \in \mathcal{S}_1$ there is a natural twisting map $b : C(X) \to C(GX)$ such that $b(x_1) = \tau(x_1)^{-1} - 1 \ \ (x_1 \in X_1)$.*

(b) *Such a natural twisting map b is unique up to natural twisting homotopy.*

(c) *For $X \in \mathcal{S}_2$ the unique map $\mathcal{B} : \Omega C(X) \to C(GX)$ in $\langle \mathrm{dgA} \rangle_1$ with $\mathcal{B}t = b$ is an H_*-equivalence.*

Recall (2.22) (a) the one-one correspondence between natural maps $\Omega C \to CG$ and natural twisting maps $C \to CG$ given by $\mathcal{B} \mapsto \mathcal{B}t$, and analogously for homotopies.

Proof. Ad (a). First b_1 is well-defined: if x_1 is degenerate then $x_1 \in \operatorname{Im} s_0$ and thus $\tau(x_1)^{-1} - 1 = 0$. Further $\varepsilon\, b_1 = 0$. Now by (2.66) (a) it suffices to show:

(i) $C(X)$ as a functor on $\mathcal{S}_1$ is degreewise representable with models $\overline{\Delta}[p]$ $(p \geqslant 0)$.

(ii) The homology of $C(GX)$ is concentrated in degree 0 when X is a model.

(iii) There is a natural map $b_2 : C_2(X) \to C_1(GX)$ with $d\, b_2 + b_1 d = m(b_1 \otimes b_1)\Delta$.

Ad (i) see (2.38) (b). — *Ad* (ii). Up to π_*-equivalence $\overline{\Delta}[p]$ is a finite bouquet of 1-spheres. Hence the homology of $G(\overline{\Delta}[p])$ is concentrated in degree 0 where it is the group algebra of a free group. — *Ad* (iii). An easy computation shows that $b_2 : x_2 \mapsto \tau(x_2)^{-1} \cdot s_0\tau(d_0 x_2)^{-1}$ has the desired property. Note that this element of $C_1(GX)$ is degenerate if x_2 is so, thus b_2 is well-defined.

Ad (b). We have to show that any two natural twisting maps extending b_1 are naturally twisting homotopic. This follows from (i)–(ii) by (2.66) (b).

Ad (c). By the Universal Coefficient Theorem we may assume $\mathsf{K} = \mathbb{Z}$. Regard the natural chain map

$$\mathrm{id} \otimes \mathcal{B} \, : \, C(X) \otimes_t \Omega C(X) \longrightarrow C(X) \otimes_b C(GX)$$

of twisted tensor products (2.24). These chain complexes are acyclic, i.e. their canonical units are H_*-equivalences, see (2.24), (4.11). Filter these chain complexes by the degree in $C(X)$ and regard the corresponding first quadrant spectral sequences. If X is 2-reduced, one checks that ∂_t, ∂_b (2.24) lower filtration by 2. Hence the E^2 terms are $H_*(X;H_*(\Omega CX))$, $H_*(X;H_*(GX))$. By Künneth (recall $\mathsf{K}=\mathbb{Z}$) and a comparison theorem [Mc2] for spectral sequences we conclude that $H_*(\mathcal{B})$ is an isomorphism. □

Remarks (4.11)

Part (a) is equivalent to [May|31.3]: There is a natural mapping T assigning to each twisting function $\theta : X \to H$ (where $X \in \mathcal{S}_1$, $H \in \langle \text{sGp} \rangle$) a twisting map $T(\theta) : C(X) \to C(H)$ such that $T(\theta)(x_1) = \theta(x_1)^{-1} - 1$. In fact, given such T put $b_X = T(\tau_X)$, and the converse follows by naturality. (*Note*: The "twisting cochains" in May's book are precisely the negatives of our twisting maps). — The twisted Eilenberg-Zilber Theorem [May|31.7] establishes a natural (filtration preserving) chain homotopy equivalence $C(X)\otimes_{T(\theta)}C(H) \simeq C(H\times_\theta X)$ extending the canonical isomorphism in degree 0. For example, since $GX \times_\tau X$ has the homotopy type of a point, $C(X)\otimes_b C(GX)$ is acyclic. Parts (a) and (c) are proved similarly in [Pr]. For a different proof of (c), see (4.76).

Corollary (4.12)

For $X \in \mathcal{S}_2$ there is a natural H_*-equivalence

$$\mathcal{B} : \Omega C(X) \xrightarrow{\sim} C(GX) \qquad \text{in } \langle \text{dgA} \rangle_1 .$$

In fact, there is a distinguished collection of such natural maps, two members of which are naturally homotopic. □

We denote by $\mathcal{B}$ any fixed member of this collection; i.e. $\mathcal{B}$ is the restriction of a natural map under $\mathcal{S}_1$ which is canonical in degree 0.

In particular $\Omega C(X)$ is the simplicial analog of the classical Adams-Hilton model for CW-complexes. Surprisingly this does not seem to be well-known.

N°4 Compatibility with (tensor) products

The result of this subsection is valid for any *natural* twisting map $b : C(X) \to C(GX)$ where $X \in \mathcal{S}_1$.

For $(X,Y) \in \mathcal{S}_1 \times \mathcal{S}_1$ regard the natural diagram:

$$(4.13)\qquad \begin{array}{ccc} CX\otimes CY & \xrightarrow{\ \mathcal{EM}\ } & C(X\times Y) \\ & & \big\downarrow b \\ \Big\downarrow b*b & \circleddash & CG(X\times Y) \\ & & \big\downarrow C(\iota) \\ C(GX)\otimes C(GY) & \xrightarrow{\ \mathcal{EM}\ } & C(GX\times GY) \end{array}$$

The marked arrows are (degree -1) twisting maps, hence so are the compositions from the top-left to the bottom-right corner. In fact, the top $\mathcal{EM}$ is in $\langle \text{dgC} \rangle_1$

while $C(\iota)$ and the bottom $\mathcal{EM}$ are in $\langle \mathrm{dgA}\rangle_0$, see (B.10). Here the sGp-map $\iota : G(X \times Y) \to GX \times GY$ is given by the twisting function $\tau_X \times \tau_Y$. Clearly ι is a π_*-equivalence, but this is not needed. For the definition of $b * b$, see Chp. 2, § 2, N° 3.

THEOREM (4.14)
There is a natural twisting homotopy $h : \mathcal{EM}\,(b * b) \simeq C(\iota)\, b\, \mathcal{EM}$.

Proof. By (2.66) (b) this is a consequence of the following three facts.

(i) $CX \otimes CY$ as a functor on $\mathcal{S}_1 \times \mathcal{S}_1$ is representable in each degree with models $(\overline{\Delta}[p], \overline{\Delta}[q])$ $(p, q \geqslant 0)$. — See (2.38) (c).

(ii) The homology of $C(GX \times GY)$ is concentrated in degree 0 when (X, Y) is a model. — Clear by the corresponding statement in the proof of (4.10) (a).

(iii) On $(CX \otimes CY)_1$ these twisting maps agree.

In order to prove the latter let $x_0 \in X_0$ be the unique 0-simplex and $y_1 \in Y_1$, then

$$\begin{aligned}
&\mathcal{EM}_0(b * b)_1(x_0 \otimes y_1) = \\
&= \eta\varepsilon(x_0) \otimes b(y_1) + b(x_0) \otimes \eta\varepsilon(y_1) && \text{since} \quad \mathcal{EM}_0 = \mathrm{id} \\
&= 1 \otimes b(y_1) \\
&= C_0(i_{GY}) b(y_1) && \text{where} \quad i_{GY} : GY = 1 \times GY \subset GX \times GY \\
&= C_0(\iota) C_0(G(i_Y)) b(y_1) && \text{where} \quad i_Y : Y = x_0 \times Y \subset X \times Y \\
&= \iota_0 b_1 C_1(i_Y)(y_1) && \text{by naturality of } b \\
&= \iota_0 b_1 (s_0 x_0 \otimes y_1) \\
&= \iota_0 b_1 \mathcal{EM}_1(x_0 \otimes y_1)\,;
\end{aligned}$$

and similarly one checks the equality on $x_1 \otimes y_0$. This proves the assertion. □

Remark. For a particular choice of b diagram (4.13) commutes strictly, see [MoPr]. One obtains immediately an equivalent result for $\mathcal{B}$, see (4.17). Compare the general discussion (B.2).

N° 5 The homotopy diagonals

We extend our result on the connection between $\Omega C(X)$ and $C(GX)$ by showing that "the" weak equivalence $\mathcal{B} : \Omega C(X) \xrightarrow{\sim} C(GX)$ is compatible with the canonical homotopy Hopf diagonals which we now explain.

PROPOSITION (4.15)
Let G be a reduced sGp. Its chain algebra $C(G)$, endowed with the homotopy diagonal $[\mathcal{AW} \circ C(\Delta)]$, is a dgHh. We thus obtain a functor $CG : \mathcal{S}_2 \to \langle \mathrm{dgHh}\rangle_1$.

Proof. Since $C(G)$ with the diagonal $\mathcal{AW} \circ C(\Delta)$ is a non-cocommutative dg Hopf algebra we only have to show that $[\mathcal{AW} \circ C(\Delta)]$ is cocommutative, i.e. the following (solid) diagram commutes (1.2) in $\mathrm{Ho}\langle \mathrm{dgA}\rangle_1$:

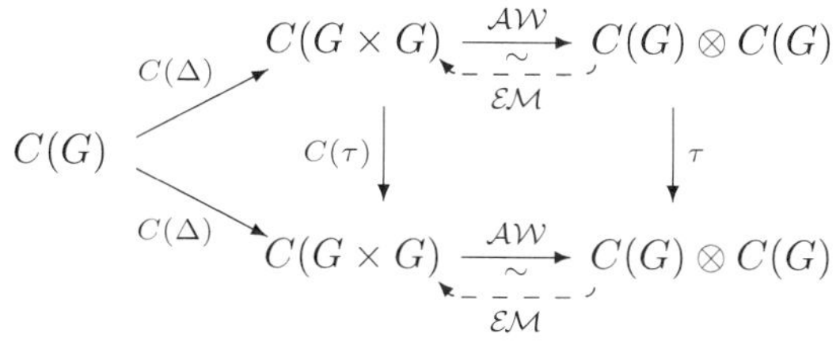

$$
\begin{array}{c}
C(G) \xrightarrow{C(\Delta)} C(G\times G) \xrightarrow[\mathcal{EM}]{\mathcal{AW}\;\sim} C(G)\otimes C(G) \\
\downarrow C(\tau) \qquad\qquad \downarrow \tau \\
C(G\times G) \xrightarrow[\mathcal{EM}]{\mathcal{AW}\;\sim} C(G)\otimes C(G)
\end{array}
$$

The triangle commutes strictly, and so does the square with the broken arrows. Since $\mathcal{AW}$ is a weak equivalence and $\mathcal{AW} \circ \mathcal{EM} = 1$ it follows that $[\mathcal{AW}]$ and $[\mathcal{EM}]$ are inverse isomorphisms in $\mathrm{Ho}\langle \mathrm{dgA} \rangle_1$ yielding the assertion. □

Theorem (4.16)
For $X \in \mathcal{S}_2$ let $\Psi = [j][\Omega(\mathcal{EM})]^{-1}[\Omega C(\Delta)]$ in $\mathrm{Ho}\langle \mathrm{dgA} \rangle_1$. This is the unique homotopy Hopf diagonal on $\Omega C(X)$, such that $\mathcal{B} : \Omega C(X) \xrightarrow{\sim} CG(X)$ is a dgHh-map.

I.e. Ψ is induced by the top row of the diagram below in $\langle \mathrm{dgA} \rangle_1$. We have to prove its commutativity up to homotopy, which implies that the induced diagram in the derived category $\mathrm{Ho}\langle \mathrm{dgA} \rangle_1$ is commutative. Here $\Omega(\mathcal{EM})$ is an H_*-equivalence by (2.28) (c). Since weak equivalences induce isomorphisms in the derived category, the uniqueness of Ψ is clear, and so is the assertion that Ψ is a homotopy *Hopf* diagonal (4.15). (It is also not difficult to show this directly using the corresponding properties of $\mathcal{EM}$ and j.)

(4.17)
$$
\begin{array}{c}
\Omega C(X) \xrightarrow{\Omega C(\Delta)} \Omega C(X\times X) \xleftarrow[\sim]{\Omega(\mathcal{EM})} \Omega(CX\otimes CX) \xrightarrow{j} \Omega C(X)\otimes \Omega C(X) \\
\mathcal{B}\Big\downarrow\sim \qquad \sim\Big\downarrow\mathcal{B} \qquad\qquad \cong \qquad\qquad \sim\Big\downarrow \mathcal{B}\otimes\mathcal{B} \\
\qquad\qquad CG(X\times X) \\
\mathrm{CG}(\Delta)\nearrow \qquad \nwarrow\sim\; C(\iota) \\
C(GX) \xrightarrow{C(\Delta)} C(GX\times GX) \xleftarrow[\mathcal{AW}]{\mathcal{EM}\;\sim} C(GX)\otimes C(GX)
\end{array}
$$

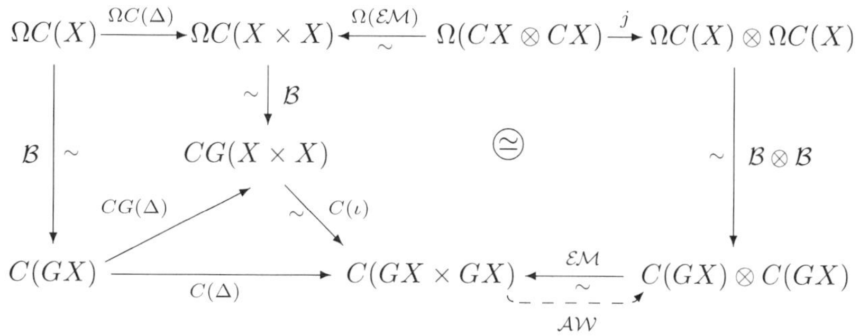

Proof. It suffices to show that the diagram above commutes in $\mathrm{Ho}\langle \mathrm{dgA} \rangle_1$. It is clear that the quadrangle and triangle on the left commute; so we are done if the (solid) hexagon starting at $\Omega(CX \otimes CX)$ (which is cofibrant!), commutes up to derivation homotopy. Let $t : CX \otimes CX \to \Omega(CX \otimes CX)$ be the universal twisting map. Now by (2.22) (a) it suffices to show that the twisting maps

$$\mathcal{EM}(\mathcal{B} \otimes \mathcal{B})jt = \mathcal{EM}(\mathcal{B} \otimes \mathcal{B})(t * t) = \mathcal{EM}(b * b)\,,$$

$$C(\iota)\,\mathcal{B}\,\Omega(\mathcal{EM})\,t = C(\iota) b\,\mathcal{EM}$$

are twisting homotopic (recall $b = \mathcal{B}t$). This holds by (4.14). □

Regard $\mathcal{S}_2$ as a homotopical category in such a way that a weak equivalence is also an H_*-equivalence. Then ΩC and thus CG preserve weak equivalences, see proof of (4.10) (c). In other words, these are fully homotopical functors. This implies that they induce (fully) derived functors between the corresponding derived categories, see Chp. 1, § 2, N° 3.

We summarize our main results as follows:

THEOREM–A (4.18)

The functors ΩC, $CG : \mathcal{S}_2 \rightrightarrows \langle \mathrm{dgHh} \rangle_1$ *are fully homotopical. There is a canonical homotopy class of weak equivalences*

$$\mathcal{B} : \Omega C(X) \xrightarrow{\sim} C(GX) \qquad \textit{in } \langle \mathrm{dgHh} \rangle_1 .$$

□

Rational equivalence (4.19)

Let $\mathsf{K} = \mathbb{Q}$ and endow $\mathcal{S}_2$ with the rational homotopy structure. We show in § 4, N° 2 that the functor ΩC in (4.18) is a homotopical equivalence (4.31). Hence Theorem–A shows that CG has the same property. This is a simplicial analog of a result of Anick.

§ 2 THE SULLIVAN MODEL

In this section K is a hereditary $\mathbb{Q}$-algebra. We adhere to the general conventions in this chapter, except that in N° 1 and N° 2 we also consider non-pointed spaces and non-augmented algebras.

N° 1 The homotopical category of commutative dg* algebras

Let $\langle \mathrm{cdg^*A} \rangle$ be the category of commutative dg* algebras over K. Recall that dg* means *positively graded with differential of degree* +1. Recall that a graded algebra A is *commutative* if $xy = (-1)^{|x|\,|y|} yx$ for $x, y \in A$. Note that for $A \in \langle \mathrm{cdg^*A} \rangle$ the (covariant) cohomology $H^*(A)$ is again a commutative graded algebra.

Free commutative algebra (4.20)

The forgetful functor $\langle \mathrm{dg^*} \rangle \leftarrow \langle \mathrm{cdg^*A} \rangle$ has a left adjoint denoted by S. In case V is free, we call $\mathsf{S}(V)$ the *free* cdg*A on V. It is clear that $\mathsf{S}(V) = \mathsf{S}(V^{\mathrm{even}}) \otimes \mathsf{S}(V^{\mathrm{odd}})$ and easy to show that $\mathsf{S}(V^{\mathrm{odd}})$ is the exterior algebra resp. $\mathsf{S}(V^{\mathrm{even}})$ the polynomial algebra, on any basis of V^{odd} resp. V^{even}.

Note that coproducts in $\langle \mathrm{cdg^*A} \rangle$ are given by the usual tensor product $\otimes$ of graded algebras.

The definition of cofibrant objects in $\langle \mathrm{cdg^*A} \rangle$ is complicated by $|d| = +1$. Not all objects which are free as algebra are allowed, since relations such as $dx = xy$ may occur. We do not give the usual "static" definition of KS-algebras, but a "dynamical" definition which is easily seen to be equivalent. For this we introduce the notion of transfinite composition.

Let I be a well-ordered set with minimal element 1 and having a maximal element χ. Let $\alpha = (X_i, \alpha_{i,j}) : I \to \mathcal{C}$ be a *convergent functor*, i.e. $\operatorname{colim}_{j \in J} \{X_j\} = X_{\sup J}$ for all $J \subset I$. Then $\alpha_{1,\chi}$ is the *transfinite composition* of the family of maps $\{\alpha_{i,i+1}\}_{i \in I}$.

Definitions (4.21)

Let $B \in \langle \mathrm{cdg^*A} \rangle$ and $b \in B$ a cocycle; let x be a variable of degree $|x| = |b| - 1 \geqslant 0$. Then $A = (B \otimes \mathsf{S}x, d)$ with $d|_B = d_B$ and $dx = b$ is a well-defined cdg*A. Such a map $B \hookrightarrow A$ is a *single attachment*. A *KS-extension* is a transfinite composition of single attachments. A *KS-algebra* is an object M such that the unit $\mathsf{K} \to M$ is a KS-extension.

THEOREM (4.22)

The category $\langle \mathrm{cdg^*A} \rangle$ *has a canonical homotopy structure: the weak equivalences are the* H^**-equivalences, the cofibrant objects are the KS-algebras, and all objects are fibrant.*

An analogous result holds in the augmented case, i.e. for $\langle \mathrm{cdg^*A} \rangle_0$. Here an object is *cofibrant*, etc. if it is so-called in $\langle \mathrm{cdg^*A} \rangle$.

The theorem as stated is actually true only if K is a field, since in general one needs cofibrant objects which are not KS-algebras. However we actually need (4.24) below which holds over any hereditary $\mathbb{Q}$-algebra.

There is a very simple natural path object in $\langle \mathrm{cdg^*A} \rangle$ resp. $\langle \mathrm{cdg^*A} \rangle_0$, see proof of (4.24). At least over a field there is also a natural cylinder object (due to Sullivan), see [B].

Proof. It is straightforward to show that $\langle \mathrm{cdg^*A} \rangle$ is a model category where a *fibration* is a surjective map, and a *cofibration* is a transfinite composition of maps of the following two types: (i) a single attachment, (ii) a surjective map whose kernel is ideal-generated by a 0-cocycle. (Notice that both can be viewed as a "killing of a cocycle" — if in (i) the cocycle is zero, a new cocycle is created.) It is clear that KS-extensions are cofibrations. Hence KS-algebras are cofibrant, and the converse is also clear (here we need that K is a field). The crucial point is in the verification of the factorization axiom (M-2) (ii). Here we need that $B \hookrightarrow B \otimes \mathsf{S}(V \oplus dV)$ is a weak equivalence. In case $B = \mathsf{K}$ this follows from the proof of (4.1) and the fact that (over a $\mathbb{Q}$-algebra) the symmetric algebra is a retract of the tensor algebra. The general case then follows by Künneth (4.5).

For general reasons $\langle \mathrm{cdg^*A} \rangle_0$ inherits model structure. For more details (in the rational case), see [BouG]. □

Definition (4.23)

Let $\langle \mathrm{cdg^*A} \rangle'_{\langle 2 \rangle}$ be the full subcategory of $\langle \mathrm{cdg^*A} \rangle_0$ consisting of degreewise flat objects A for which $H^*(A)$ is 2-reduced, of finite type and free in degree 2.

Let $\mathbf{M}'_2$ be the full subcategory of 2-reduced KS-algebras *of finite type*.

COROLLARY (4.24)

The category $\langle \mathrm{cdg^*A} \rangle'_{\langle 2 \rangle}$ *has a canonical right homotopy structure: the weak equivalences are the* H^**-equivalences, the cofibrant objects are the* 2*-reduced KS-algebras of finite type, and all objects are fibrant.*

The homotopy relation is given by an explicit natural path object.

Proof. Recall that $\langle \mathrm{cdg^*A} \rangle_0$ is a model category. A natural path object is given by $PA = \mathsf{K} \oplus (\overline{A} \otimes \mathsf{S}(t, dt))$ with obvious maps $p_0, p_1 : PA \rightrightarrows A$ and $s : A \to PA$, cf. [BouG]. One verifies that any object $A \in \langle \mathrm{cdg^*A} \rangle'_{\langle 2 \rangle}$ has a cofibrant model $M \xrightarrow{\sim} A$ where $M \in \mathbf{M}'_2$. Hence $\langle \mathrm{cdg^*A} \rangle'_{\langle 2 \rangle}$ is a right homotopical subcategory by

(1.16). It is clear that we may choose the cofibrant objects as stated. □

Remark (4.25)

Moreover, any cofibrant object M has a cofibrant path object. This is clear by the method of factorization according to (M-2) (ii) in the proof of (4.22). Hence $\mathbf{M}_2'$ is a right homotopical subcategory of $\langle \mathrm{cdg^*A}\rangle'_{\langle 2\rangle}$ and thus localizing, see (1.12). It is also easily seen that $\langle \mathrm{cdg^*A}\rangle'_{\langle 2\rangle}$ is a fibration category where the cofibrant objects are the KS-algebras in this category.

The following is of interest in its own right. A proof is given later (4.52) using some of our main results. Recall that $\langle \mathrm{dg^*A}\rangle_0$ has a canonical homotopy structure, where $(\simeq)$ is given by derivation homotopy, cf. (4.4) (c). For the notion of equivalence of maps, see (1.2) and (1.25) (a).

THEOREM (4.26)

Two maps in $\langle \mathrm{cdg^*A}\rangle'_{\langle 2\rangle}$ *are equivalent iff they are so in* $\langle \mathrm{dg^*A}\rangle_0$. □

This shows that the homotopy relation in $\langle \mathrm{cdg^*A}\rangle'_{\langle 2\rangle}$ may be described using free dg*A-models and derivation homotopy. One should be able to create a direct proof which makes no use of the finite type assumption (at least over a field).

N° 2 The Sullivan cofunctor and Stokes' map

We sketch the construction of the *Sullivan cofunctor*

$$A^* : \mathcal{S} \longrightarrow \langle \mathrm{cdg^*A}\rangle \,. \tag{4.27}$$

Let $A_n \in \langle \mathrm{cdg^*A}\rangle$ be the algebra of *polynomial forms* on the standard geometric n-simplex

$$\Delta_n = \{(x_0, \dots, x_n) \in \mathbb{R}^{n+1} | x_i \geqslant 0, \textstyle\sum x_i \leqslant 1\}\,.$$

In more detail, we define

$$\begin{aligned} A_n &= \text{polynomial forms on } \mathbb{R}^{n+1}/\mathrm{Ann}(\Delta_n) \\ &= \mathsf{S}(x_0, \dots, x_n, dx_0, \dots, dx_n)/(1 - \textstyle\sum_{i=0}^n x_i, \sum_{i=0}^n dx_i) \end{aligned}$$

where $\mathrm{Ann}(\Delta_n)$ is the ideal consisting of all polynomial forms which are zero on Δ_n. Of course $|x_i| = 0$, $|dx_i| = 1$. Thus $A_0 = \mathsf{K}$. The A_n form a simplicial object A over $\langle \mathrm{cdg^*A}\rangle$ in an evident way. It is not very difficult to show that A satisfies the conditions (i) and (ii) of (B.29). Hence the cofunctor $A^*(X) = \mathcal{S}(X, A)$ is representable and acyclic on models rel η. In other words A^* is a cohomology theory. Recall that we may identify $A^*(\Delta[n]) = A_n$. The following is an application of (B.31).

THEOREM (4.28)

There is a unique natural cochain map (Stokes' map)

$$\textstyle\int = \int_X : A^*(X) \longrightarrow C^*(X)\,.$$

It is an algebra map in degree 0 *and a cochain homotopy equivalence inducing a graded algebra isomophism on cohomology.* □

Remark (4.29)

One can show that Stokes' map is given by the formula

$$\langle \textstyle\int \omega, x\rangle = \int_{\Delta[n]} \omega(x) \qquad (\omega \in A^n(X)\,,\ x \in X^n)$$

where on the right-hand side we mean true integration; see [BouG|p. 5 ff]. Since Stokes' map is *not* multiplicative, it follows from (4.28) that there cannot exist *any* natural dg* *algebra* map $A^* \to C^*$. In §3 we show that Stokes' map *is* however strongly homotopy multiplicative.

If $x_0 : * \to X$ is a pointed space, then $A^*(x_0) : A^*(X) \to A^*(*) = \mathsf{K}$ is an augmented cdg*A. This defines the cofunctor

(4.30) $$A^* : \mathcal{S}_0 \longrightarrow \langle \mathrm{cdg^*A}\rangle_0\,.$$

Observe that Stokes' map is compatible with augmentation and unit. It is clear by (4.28) that a map f in $\mathcal{S}_0$ (or $\mathcal{S}$) is an H^*-equivalence iff $A^*(f)$ is so.

Let $\mathcal{S}_2$ be endowed with its rational homotopy structure, and let $\mathcal{S}_2' \subset \mathcal{S}_2$ be the homotopical subcategory (1.13) consisting of spaces X of finite rational type, i.e. $H^*(X;\mathbb{Q})$ is finite dimensional in each degree. Consider the restriction

(4.31) $$A^* : \mathcal{S}_2' \longrightarrow \langle \mathrm{cdg^*A}\rangle_{\langle 2\rangle}'$$

which preserves weak equivalences. This restriction exists since $A^*(X)$ is always degreewise flat by (B.24). Given $X \in \mathcal{S}_2'$ a cofibrant model $\ell : M \xrightarrow{\sim} A^*(X)$ is said to be a *KS-model* of X, cf. (4.24).

We close this introduction to Sullivan's theory by mentioning some further properties of A^*. From these we only need however that this cofunctor carries colimits to limits.

Geometric realization (4.32)

The cofunctor A^* (4.27) has an adjoint $\|-\| : \langle \mathrm{cdg^*A}\rangle \to \mathcal{S}$, see App. B, N° 4. By definition an n-simplex of $\|A\|$ is just a map $A \to A_n$ and the simplicial operators are obvious. Hence A^* and $\|-\|$ carry colimits to limits. Using that A^p is a contractible Kan complex, one easily sees that A^* carries cofibrations (injective maps) to fibrations (surjective maps). Hence $(A_*, \|-\|)$ is an adjoint homotopical pair (A.16).

The Sullivan equivalence (4.33)

Let $\mathsf{K} = \mathbb{Q}$. The principal result of Sullivan's theory states that A^* (4.31) is a homotopical equivalence (1.31). Using geometric realization one can prove this as follows. It is easily seen that $\|\mathsf{S}(x)\|$ is an Eilenberg-MacLane space $K(\mathbb{Q}, |x|)$. Hence the realization of $M \in \mathbf{M}_2'$ is a rational Kan complex of finite rational type. Hence $\|-\|$ restricted to $\mathbf{M}_2'$ is adjoint to A^* (4.31). By the Adjoint Functor Theorem (1.32) we are reduced to show that the adjunction map $M \to A^*\|M\|$ is an H^*-equivalence. The key observation is that the obvious increasing algebra filtration of M corresponds geometrically to the Postnikov tower of $\|M\|$. For details see [BouG], [FU]. For proofs that make now use of geometric realization [B], [GrMg], [Hal].

N° 3 Extension of Stokes' map

From now on we work with *pointed* spaces and A^* is the Sullivan cofunctor (4.30).

If X is a pointed space, the bar construction of $A^*(X) = A$ is defined: $BA = \mathsf{T}'(\mathrm{s}^{-1}\overline{A})$ as a graded coalgebra. Note that in general BA is *not positive* (nor negative) — even if X is 2-reduced; cf. (2.21). Let $BA = \bigoplus_{n\geqslant 0} B_n A$ be the canonical decomposition, and i_n the inclusion of $B_n A = (\mathrm{s}^{-1}\overline{A})^{\otimes n}$.

Put $w_1 = -\int t_1 : B_1 A^*(X) \to C^*(X)$ where $t_1 = t\, i_1$ and $t : BA^*(X) \to A^*(X)$ the universal twisting map. Then $|w_1| = +1$, $D(w_1) = 0$ and $\varepsilon\, w_1 = 0$. Using notation (2.5) we may also write $w_1 = -\langle \int \rangle$, where $\int : \overline{A} \to C$.

THEOREM (4.34)

(a) *For $X \in \mathcal{S}_0$ there is a natural twisting map*

$$w : BA^*(X) \longrightarrow C^*(X) \qquad \text{with} \quad w\, i_1 = w_1 .$$

(b) *Two such natural twisting maps are naturally twisting homotopic.*

(c) *The unique map $\mathcal{W} : BA^*(X) \longrightarrow BC^*(X) \in \langle \mathrm{DG^*C}\rangle_0$ with $t\mathcal{W} = w$ is an H^*-equivalence.*

Proof. *Ad* (a). This is a consequence, by (2.61) (a), of the following three facts:

(i) C^* as a cofunctor on $\mathcal{S}_0$ is degreewise representable with models $\Delta[p]^+$ ($p \geqslant 0$), where $X^+ = X \sqcup *$ with $*$ as basepoint ($X \in \mathcal{S}$).

(ii) BA^* is Σ-decomposable as a cofunctor on $\mathcal{S}_0$, and $H^*(B_n A^*(M), d) = \mathrm{s}^{-n}\mathsf{K}$ if $M = \Delta[p]^+$ is a model.

(iii) The relation $w_1 \cup w_1 = w_1 \delta_2$ holds on $(B_2 A)^{-2}$ where $A = A^*(X)$.

Ad (i). An application of (2.37).

Ad (ii). The first assertion is clear by definition of the bar construction. The internal differential d of $B_n A^*$ is the tensor product differential induced from that of A^*. Now $B_n A^*(X^+) = \mathrm{s}^{-n}(A^*(X)^{\otimes n})$ and since $A^*(M)$ is acyclic rel η the second assertion follows. Note that $A^*(X^+) = A^*(X) \times \mathsf{K}$ with augmentation $\varepsilon = \mathrm{pr}_2$, since A^* carries coproducts to products (4.32).

Ad (iii). This follows from the fact that Stokes' map is an algebra map in degree 0, cf. (2.62).

Ad (b). This follows, by (2.61) (b), from (i)–(ii).

Ad (c). The coaugmental filtration of the bar construction is cocomplete (2.15) and bounded below, thus bicomplete. Hence by [EM] it suffices to show that $E_1\mathcal{W}$ is isomorphic. One readily checks that $E_0^{*,k}(\mathcal{W})$ is the map $\langle \overline{w}_1 \rangle^{\otimes k} : (\mathrm{s}^{-1}\overline{A})^{\otimes k} \to (\mathrm{s}^{-1}\overline{C})^{\otimes k}$ which is an H^*-equivalence by Künneth (4.5) and since $\langle \overline{w}_1 \rangle = -\mathrm{s}^{-1}\overline{\int}$. □

Remarks (4.35)

(a) One may regard $w = \sum w_n$ as a "system of higher homotopies" whose existence means that Stokes' map is strongly homotopy multiplicative, cf. (2.62). Compare the discussion in [BouG | § 3] where (4.34) (a) is proved similarly.

(b) The homotopy theoretic meaning of the theorem above (at least if X is 1-connected) is essentially that $A^*(X)$ and $C^*(X)$ are weakly equivalent in $\langle \mathrm{dg}^*\mathrm{A}\rangle_0$, cf. (2.29).

N° 4 Compatibility with (tensor) products

The result of this subsection is valid for any natural twisting map $w : BA^*(X) \to C^*(X)$. We leave to the reader the formulation and deduction of the corresponding result on $\mathcal{W}$.

For $(X, Y) \in \mathcal{S}_0 \times \mathcal{S}_0$ regard the natural diagram

(4.36)
$$\begin{array}{ccccc} BA^*(X) \otimes BA^*(Y) & \xrightarrow{\tilde{\jmath}} & B(A^*X \otimes A^*Y) & \xrightarrow{B(\mu)} & BA^*(X \times Y) \\ {\scriptstyle w*w}\downarrow & & \cong & & \downarrow{\scriptstyle w} \\ C^*(X) \otimes C^*(Y) & \xrightarrow{\phi} & (CX \otimes CY)^* & \xleftarrow{\mathcal{EM}^*} & C^*(X \times Y) \end{array}$$

where the marked arrows are (degree $+1$) twisting maps, the top row lies in $\langle \mathrm{DG}^*\mathrm{C}\rangle_0$ and the bottom row in $\langle \mathrm{dg}^*\mathrm{A}\rangle_0$. For the definition of $\tilde{\jmath}$ see (2.26).

If C is chain complex we regard here $C^* = \{\mathrm{Hom}(C_n, \mathsf{K}), \delta^n\}$ with differential $\delta^n(\alpha^n) = \alpha^n d_{n+1}$. The natural map ϕ is then defined by $\phi(\alpha \otimes \beta)(x \otimes y) = \alpha(x)\beta(y)$. It is this map which is used to define the algebra structure of C^* if C is a coalgebra. This is of course standard, however there is an alternative choice of signs in the definition of δ and ϕ which for some purposes is more useful, see N° 5.

The natural map

(4.37)
$$\mu : A^*X \otimes A^*Y \longrightarrow A^*(X \times Y)\,, \qquad \mu(\omega \otimes \varphi) = \mathrm{pr}_X^*(\omega) \cdot \mathrm{pr}_Y^*(\varphi)$$

is a map in $\langle \mathrm{cdg}^*\mathrm{A}\rangle_0$ where in the definition we use that $\otimes$ is the coproduct in this category. Notice that μ for A^* is the analog of $\mathcal{AW}^*$ for C^*. Relating these monoidal transformations via Stokes' map shows that μ is an H^*-equivalence.

THEOREM (4.38)
For $(X, Y) \in \mathcal{S}_0 \times \mathcal{S}_0$ there is a natural twisting homotopy $h : \mathcal{EM}^* w B(\mu)\tilde{\jmath} \simeq \phi(w * w)$.

Proof. This follows, again by (2.61) (b), from the following three facts:

(i) $(CX \otimes CY)^*$ as a cofunctor on $\mathcal{S}_0 \times \mathcal{S}_0$ is degreewise representable with models $(\Delta[p]^+, \Delta[q]^+)$ $(p, q \geqslant 0)$.

(ii) $BA^*(X) \otimes BA^*(Y)$ is Σ-decomposable as a cofunctor on $\mathcal{S}_0 \times \mathcal{S}_0$. Moreover $H^*([BA^*(M) \otimes BA^*(N)]_n, d)$ is concentrated in degree n if (M, N) is a model.

(iii) The two twisting maps in question agree on $[BA^*(X) \otimes BA^*(Y)]_1$.

Ad (i). Using standard facts, this is an application of (2.37). Note that $\mathcal{S}_0 \times \mathcal{S}_0 = (\mathcal{S} \times \mathcal{S})^{(*,*)}$ by which we identify $(X^+, Y^+) = (X, Y)^+$ for $(X, Y) \in \mathcal{S} \times \mathcal{S}$.

Ad (ii). The required decomposition of $BA^*(X) \otimes BA^*(Y)$ is given by $[BA^*(X) \otimes BA^*(Y)]_n = \bigoplus_{i+j=n} B_iA^*(X) \otimes B_jA^*(Y)$, cf. (2.60). With respect to internal differentials, this is an identity of cochain complexes. Since $H^*(B_iA^*(M), d) = \mathrm{s}^{-i}(\mathsf{K})$ if $M = \Delta[p]^+$ (proof of (4.34) (a)), the second assertion follows.

Ad (iii). We have to show that the twisting maps agree on elements $\mathrm{s}^{-1}\omega \otimes 1$ where $\omega \in \overline{A}^*(X)$ resp. $1 \otimes \mathrm{s}^{-1}\phi$ where $\phi \in \overline{A}^*(Y)$. We only exhibit the verification of the former which is actually sufficient since the maps in (4.36) commute with interchange isomorphism. By naturality of $\tilde{\jmath}$ we have

$$B(\mu)\tilde{\jmath}(\mathrm{s}^{-1}\omega \otimes 1) = B(\mu)(\mathrm{s}^{-1}(\omega \otimes 1)) = \mathrm{s}^{-1}\mu(\omega \otimes 1)\,.$$

Using that on such elements (of lower degree 1) w is given by $-\int$ we compute ...

$$\begin{aligned}
\mathcal{EM}^* w_{X\times Y} B(\mu)\tilde{\jmath}(\mathrm{s}^{-1}\omega \otimes 1) &= \mathcal{EM}^* w_{X\times Y}(\mathrm{s}^{-1}\mu(\omega \otimes 1)) \\
&= -\mathcal{EM}^* \textstyle\int_{X\times Y} \mu(\omega \otimes 1) = -\mathcal{EM}^* \int_{X\times Y} \mathrm{pr}_X^*(\omega) \\
&= -\mathcal{EM}^* C^*(\mathrm{pr}_X) \textstyle\int_X(\omega) = -\pi_1^* \int_X(\omega) \\
&= -\phi(\textstyle\int_X(\omega) \otimes 1) \qquad = \phi(w_X(\mathrm{s}^{-1}\omega) \otimes 1) \\
&= \phi(w_X * w_Y)(\mathrm{s}^{-1}\omega \otimes 1)\,.
\end{aligned}$$

Here $\pi_1 = \mathrm{id} \otimes \varepsilon : CX \otimes CY \to CX$. We use that for $x \otimes y \in CX \otimes CY$

$$\langle \pi_1^* \textstyle\int_X(\omega),\, x \otimes y\rangle = \langle \int_X(\omega),\, x\,\varepsilon(y)\rangle = \langle \phi(\int_X(\omega) \otimes 1),\, x \otimes y\rangle$$

since $1 \in C^*(Y)$ is the augmentation of $C(Y)$. This completes the proof of the desired relation. □

N° 5 Dualization

Since the target of Stokes' map resp. of w in (4.34) is the (functional) dual of a dg coalgebra, it gives rise to an adjoint map. We use this elementary but crucial fact to connect the dual DG algebra of $BA^*(X)$ to the dg algebra $\Omega C(X)$. This suggests the definition of a loop cofunctor $F^* : \langle \mathrm{cdg^*A}\rangle_0' \to \langle \mathrm{dgA}\rangle_1$.

Let $\# : \langle \mathrm{DG}^*\rangle \rightleftarrows \langle \mathrm{DG}\rangle : {}^\vee$ be the functional dualization cofunctors $\mathrm{Hom}(-, \mathsf{K})$ and let $\phi : \#E \otimes \#F \to \#(E \otimes F)$ be the obvious map, cf. proof of (4.40). We use here Chp. 2, § 3, N° 3 — example (2.57) (b).

In the following lemma, E and C are degreewise flat complexes. Recall that over a hereditary ring flat means torsion-free. Hence the functional dual of a flat module is flat.

LEMMA (4.39)

(a) *If a map f in $\langle \mathrm{DG}^*\rangle$ is an H^*-equivalence, then $\#f$ is an H_*-equivalence.*

(b) *If $f : E \to C^\vee \in \langle \mathrm{DG}^*\rangle$ is an H^*-equivalence and $H_*(C)$ is of finite type, then $f' : C \to \#E$ is an H_*-equivalence.*

(c) *If E is free of finite type, then ϕ is an isomorphism. If $H^*(E)$ is of finite type, then ϕ is an H_*-equivalence.*

Proof. *Ad* (a). An application of the Universal Coefficient Theorem for cohomology which in case K a field tells us that $H_*(\#E) = \#H^*(E)$. — *Ad* (b). We have $f' = \#f \circ j_C$ where $j_C : C \to \#(C^\vee)$ is the adjunction map. By assumption there is an H_*-equivalence $M \xrightarrow{\sim} C$ with M free of finite type. Since clearly j_M is an isomorphism, j_C is an H_*-equivalence yielding the assertion by the first part. — The proof of (c) is similar; cf. [Sp]. □

Note that $\#A^*(X)$ is *not* a coalgebra in general, but of course it has a canonical counit and (if X is pointed) coaugmentation.

PROPOSITION (4.40)

There is a unique natural counit-preserving chain map $\rho : C(X) \to \#A^*(X)$. It is coaugmented if X is pointed. Further ρ is an H_*-equivalence if X is of finite rational type. □

Proof. Of course (4.28) remains true if $C^*(X)$ is replaced by $C(X)^\vee$. Recall that these coincide up to different signs in the definition of the multiplication and the differential: in $C(X)^\vee$ we have $\langle \delta, \alpha\rangle = -(-1)^{|\alpha|}\alpha d$ and $\langle \alpha\cdot\beta, x\rangle = \langle \phi(\alpha\otimes\beta), \Delta(x)\rangle$ where Δ is the $\mathcal{AW}$-diagonal and $\phi(\alpha\otimes\beta)(x\otimes y) = (-1)^{|\beta||x|}\alpha(x)\beta(y)$; in the definition of $C^*(X)$ the signs are omitted. — With this replacement also Stokes' map gets different (and somewhat weird) signs, but now we can put $\rho = \int'$ and the desired properties of ρ are clear by (4.39). (Except for the last assertion one can also give an acyclic model argument, dual to the proof of (4.28).) □

Put $r^1 = -\#(t_1)\circ\rho : C(X) \to \#(B_1A^*(X))$. This degree -1 map satisfies $D(r^1) = 0$ and $r^1\eta = 0$. Hence we may regard r^1 as a map on $\widetilde{C}(X)$ and get the chain map $\langle r^1\rangle : s_{-1}\widetilde{C}(X) \to \#(B_1A^*(X))$, which is an H_*-equivalence if X is of finite rational type by (4.40). For notation see (2.5).

Note that $\#(BA^*) = \prod_{n\geqslant 0}\#(B_nA^*)$. Let $p^n = \#(i_n)$ be the projection onto the n-th factor.

THEOREM (4.41)

(a) *For $X \in \mathcal{S}_0$ there is a natural twisting map*

$$r : C(X) \longrightarrow \#BA^*(X) \qquad \textit{with} \quad p^1r = r^1 .$$

(b) *Two such natural twisting maps are naturally twisting homotopic.*

(c) *The unique map $\mathcal{R} : \Omega C(X) \to \#BA^*(X) \in \langle\mathrm{DGA}\rangle_0$ with $\mathcal{R}t = r$ is an H_*-equivalence if $X \in \mathcal{S}_2'$.*

Proof. With the minor modification described in the proof of (4.40), parts (a)–(b) are equivalent to the corresponding statements in (4.34): $r^1 = w_1'$, $r = w'$, etc. One can also apply the dual acyclic model theorem for twisting maps since $\#(BA^*)$ is Π-decomposable, see (2.65).

Ad (c). Put $B = BA^*(X)$. Filter $\#B$ by the decreasing filtration $F^n = \prod_{k\geqslant n}\#(B_k)$ which is dual to the coaugmental filtration of B (2.15), i.e. $F^n = \#(B/F_nB)$. Hence the two components of the differential of $\#B$ ($\#d$ and $\#\delta$) preserve filtration and $\#\delta$ raises filtration by one. — Put $C = CX$. Filter ΩC by the augmental filtration $F^n = \bigoplus_{k\geqslant n}(s_{-1}\widetilde{C})^{\otimes n}$. Again the two components of the differential of ΩC (d and ∂) preserve filtration and ∂ raises filtration by one. Let $i^n : (s_{-1}\widetilde{C})^{\otimes n} \subset \Omega C$ be the inclusion. Regard r as a map $\widetilde{C} \to \#B$. Clearly $p^k\mathcal{R}i^n$ is the composite

$$(*) \qquad (s_{-1}\widetilde{C})^{\otimes n} \xrightarrow{\langle r\rangle^{\otimes n}} (\#B)^{\otimes n} \xrightarrow{\phi} \#(B^{\otimes n}) \xrightarrow{\#\Delta^{(n)}} \#B \xrightarrow{\#i_k} \#(B_k)\,.$$

Note that $\#(\Delta^{(n)})\phi$ is the iterated multiplication on $\#B$. By (2.52) the adjoint of the graded map $(*)$ is the composite

$$(*)' \quad B_k \xrightarrow{\Delta^{(n)}} \bigoplus_{\Sigma k_i = k} B_{k_1} \otimes \ldots \otimes B_{k_n} \xrightarrow{\langle r \rangle'^{\otimes n}} ((s_{-1}\widetilde{C})^\vee)^{\otimes n} \xrightarrow{\phi} ((s_{-1}\widetilde{C})^{\otimes n})^\vee .$$

In case $k < n$ this is zero, because $\langle r \rangle'$ is zero on B_0. Thus $\mathcal{R}$ is filtration preserving. For the same reason, in case $k = n$ only the component $B_n \xrightarrow{\simeq} B_1 \otimes \ldots \otimes B_1$ of $\Delta^{(n)}$ makes a contribution in $(*)'$. Since clearly $\langle r \rangle'|_{B_1} = \langle r^1 \rangle'$ we conclude that $\mathrm{gr}^n\mathcal{R}$ is the composite

$$(s_{-1}\widetilde{C})^{\otimes n} \xrightarrow{\langle r^1 \rangle^{\otimes n}} \#(B_1)^{\otimes n} \xrightarrow{\phi} \#(B_1^{\otimes n}) = \#(B_n) .$$

Considering the spectral sequences of those filtrations, we have thus determined $E^0\mathcal{R}$. Note that on E^0C resp. $E^0(\#B)$ only the 'internal' differential d resp. $\#d$ is visible. Clearly the filtration of $\#B$ is bicomplete, and since X is 2-reduced the same is true of ΩC. Hence ([EM]) for $\mathcal{R}$ to be an H_*-equivalence it suffices that $E^1\mathcal{R}$ is isomorphic. For this it suffices that $\langle r^1 \rangle$ and ϕ are H_*-equivalences which is the case since X is of finite rational type. □

Definition (4.42)
The *loop cofunctor* $F^* : \langle \mathrm{cdg}^*\mathrm{A} \rangle'_{\langle 2 \rangle} \to \langle \mathrm{dgA} \rangle_1$ is given by $F^*(A) = R_1(\#BA)$. Here the functor $R_1 : \langle \mathrm{DGA} \rangle_0 \to \langle \mathrm{dgA} \rangle_1$ sends A to its sub-DGA: $\mathsf{K} \oplus Z_1 A \oplus A_{>1}$.

One checks that R_1 is right adjoint-left inverse to the embedding, R_1 preserves H_*-equivalences and $R_1(A) \hookrightarrow A$ is an H_*-equivalence if A is *connected*, i.e. $H_{<1}(A) = \mathsf{K}$. Compare proof of (4.1).

LEMMA (4.43)
(a) *F^* carries H^*-equivalences to H_*-equivalences.*
(b) *F^*A is flat and $H_*(F^*A)$ is of finite type.*
(c) *The inclusion $F^*(A) \hookrightarrow \#BA$ is an H_*-equivalence.*

Proof. *Ad* (a) see (2.28), (4.39) (a).— *Ad* (b). The former is clear since the coefficient ring is hereditary. Choose a cofibrant model $M \xrightarrow{\simeq} A \in \langle \mathrm{cdg}^*\mathrm{A} \rangle'_{\langle 2 \rangle}$. Then $F^*A \xrightarrow{\simeq} F^*M$ and since clearly F^*M is of finite type this proves the assertion.— *Ad* (c). We have $\#BA \xrightarrow{\simeq} \#BM = F^*M$, thus $\#BA$ is connected. □

Observe that the image of $\mathcal{R} : \Omega C(X) \to \#BA^*(X)$ in (4.41) (c) is contained in $F^*A^*(X)$ provided X is 2-reduced (and the same holds for any derivation homotopy from $\mathcal{R}$ to another individual). Whence the following which (by § 1) is a geometrical justification for calling $F^*(A)$ the 'loop algebra' of A.

COROLLARY (4.44)
For $X \in \mathcal{S}'_2$ there is a natural H_*-equivalence

$$\mathcal{R} : \Omega C(X) \xrightarrow{\simeq} F^*A^*(X) \qquad \text{in } \langle \mathrm{dgA} \rangle_1 .$$

In fact, there is a distinguished collection of such natural maps, two members of which are naturally homotopic. □

We denote by $\mathcal{R}$ any fixed member of this collection.

N° 6 The homotopy diagonals

We construct a homotopy Hopf diagonal on the loop algebra $F^*(A)$ where $A \in \langle \mathrm{cdg^*A} \rangle'_{\langle 2 \rangle}$. Then we extend our result (4.44) on the connection between $\Omega C(X)$ and $F^*A^*(X)$ by showing that $\mathcal{R}$ is a dgHh-map.

$F^*(A)$ as a Hopf algebra up to homotopy (4.45)

Let $A \in \langle \mathrm{cdg^*A} \rangle'_{\langle 2 \rangle}$. Then, because A is commutative, the shuffle multiplication $\nu : BA \otimes BA \to BA$ (2.27) gives BA the structure of a (commutative, associative) DG Hopf coalgebra.

We define $\Phi_A : F^*A \to F^*A \otimes F^*A \in \mathrm{Ho}\langle \mathrm{dgA} \rangle_1$ as follows: $\Phi_A = [\phi\,\iota^{\otimes 2}]^{-1} \circ [\#\nu\,\iota]$ is the image in $\mathrm{Ho}\langle \mathrm{dgA} \rangle_1$ of the middle row of (4.46).

In this diagram we choose a cofibrant model $\ell : M \xrightarrow{\sim} A$ in $\langle \mathrm{cdg^*A} \rangle'_{\langle 2 \rangle}$. Then all vertical maps are H_*-equivalences by (4.43). Further BM is of finite type and so the top ϕ is an isomorphism. We conclude that $\phi\,\iota^{\otimes 2}$ is an H_*-equivalence, thus our definition of Φ makes sense.

$$
\begin{array}{rcccc}
\boldsymbol{F^*M} = & \#BM & \xrightarrow{\#\nu} & \#(BM \otimes BM) & \overset{\phi}{=\!=\!=} & \#BM \otimes \#BM \\
& \sim \uparrow \#B(\ell) & & \sim \uparrow \#(B(\ell) \otimes B(\ell)) & & \sim \uparrow \#B(\ell) \otimes \#B(\ell) \\
\boldsymbol{F^*A} = & R_1(\#BA) & \xrightarrow{\#\nu\,\iota} & R_1(\#(BA \otimes BA)) & \underset{\sim}{\xleftarrow{\phi\,\iota^{\otimes 2}}} & R_1(\#BA) \otimes R_1(\#BA) \\
& \sim \downarrow \iota & & \sim \downarrow \iota & & \sim \downarrow \iota^{\otimes 2} \\
(4.46) \quad & \#BA & \xrightarrow{\#\nu} & \#(BA \otimes BA) & \xleftarrow{\phi} & \#BA \otimes \#BA
\end{array}
$$

PROPOSITION (4.47)

For $A \in \langle \mathrm{cdg^*A} \rangle'_{\langle 2 \rangle}$ the loop algebra $F^*(A) = R_1(\#BA)$ has the canonical homotopy Hopf diagonal Φ_A.

For $M \in \mathbf{M}'_2$ we have a canonical dgH-isomorphism $F^*M = \Omega(\#M)$ and Φ_M is the homotopy class of the shuffle diagonal.

Actually it is sufficient that M is 2-reduced and degreewise free of finite type.

Proof. It is clear that $(F^*M, \phi^{-1}\#\nu) \in \langle \mathrm{dgH} \rangle_1$. Thus $(F^*M, \Phi_M) \in \langle \mathrm{dgHh} \rangle_1$. From $F^*(\ell) : F^*A \xrightarrow{\sim} F^*M$ it follows that also $(F^*A, \Phi_A) \in \langle \mathrm{dgHh} \rangle_1$. For the second assertion apply (4.39) (c) and check that the diagonal $\phi^{-1}\#\nu$ of $\#BM = \Omega(\#M)$ is just the shuffle diagonal. □

Recall that $\Omega C(X)$ has a canonical homotopy Hopf diagonal Ψ, see (4.16). Here is the desired extension of (4.44):

THEOREM (4.48)

For $X \in \mathcal{S}'_2$ the natural dgA-map $\mathcal{R} : \Omega C(X) \xrightarrow{\sim} F^*A^*(X)$ is compatible with the canonical homotopy diagonals, i.e. $\mathcal{R}$ is a map in $\langle \mathrm{dgHh} \rangle_1$.

Proof. We have to show that $\Phi \circ [\mathcal{R}] = ([\mathcal{R}] \otimes [\mathcal{R}]) \circ \Psi$ in $\mathrm{Ho}\langle \mathrm{dgA}\rangle_1$. Regard the following diagram in $\langle \mathrm{DGA}\rangle_0$ and note that the top row coincides with the bottom row of (4.46) when $A = A^*(X)$.

(4.49)

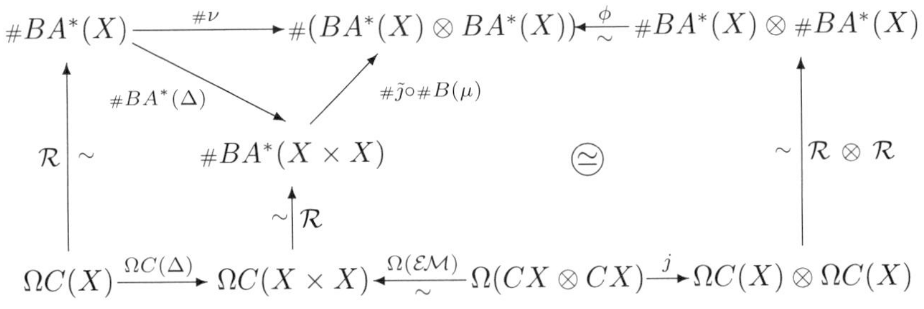
$\#BA^*(X)$
$\#\nu$
$\#(BA^*(X) \otimes BA^*(X))$
ϕ
$\#BA^*(X) \otimes \#BA^*(X)$
$\#BA^*(\Delta)$
$\#\tilde{\jmath} \circ \#B(\mu)$
$\mathcal{R}$
$\#BA^*(X \times X)$
$(\simeq)$
$\mathcal{R} \otimes \mathcal{R}$
$\mathcal{R}$
$\Omega C(X)$
$\Omega C(\Delta)$
$\Omega C(X \times X)$
$\Omega(\mathcal{EM})$
$\Omega(CX \otimes CX)$
j
$\Omega C(X) \otimes \Omega C(X)$

The top left triangle commutes by $\nu = B(m)\tilde{\jmath}$ (2.27) and the obvious factorization $m = A^*(\Delta) \circ \mu$ in $\langle \mathrm{cdg^*A}\rangle_0$ of the multiplication of $A^*(X)$; cf. (4.37). The left quadrangle commutes since $\mathcal{R}$ is natural.

Suppose there is a derivation homotopy H for the hexagonal subdiagram. Then apply R_1 to (4.49), which leaves the bottom row unchanged. Clearly $\mathrm{Im}(\mathcal{R} \otimes \mathcal{R}) \subset F^*A^*(X) \otimes F^*A^*(X)$ by which we may replace thus the top right corner, and obtain the middle row of (4.46). Clearly H is a derivation homotopy for the reduced hexagon as well. Hence the image in $\mathrm{Ho}\langle \mathrm{dgA}\rangle_1$ of the reduced diagram is commutative. As $[\Omega(\mathcal{EM})]$ and $[\phi\, \iota^{\otimes 2}]$ are isomorphisms this yields the result.

Hence it remains to show that the hexagon commutes up to derivation homotopy, or equivalently (2.22) (a), that the two twisting maps

$$\#\tilde{\jmath} \circ \#B(\mu) \circ \mathcal{R} \circ \Omega(\mathcal{EM}) \circ t = \#(B(\mu)\tilde{\jmath}) \circ r \circ \mathcal{EM}\,,$$
$$\phi \circ (\mathcal{R} \otimes \mathcal{R}) \circ j \circ t = \phi(\mathcal{R} \otimes \mathcal{R})(t * t) = \phi\,(r * r)$$

are homotopic. Again equivalently (2.55) we show that the *adjoint* twisting maps are homotopic, but clearly these are given by diagram (4.36). Hence the proof is complete by Theorem (4.38). □

Let $\langle \mathrm{dgHh}\rangle_1'$ be the homotopical subcategory of $\langle \mathrm{dgHh}\rangle_1$ consisting of all objects with homology of finite type. We consider the composite functor F^*A^* :

$$\mathcal{S}_2' \xrightarrow{A^*} \langle \mathrm{cdg^*A}\rangle_{\langle 2\rangle}' \xrightarrow{F^*} \langle \mathrm{dgHh}\rangle_1' \tag{4.50}$$

which preserves weak equivalences. We may summarize our main results (4.48), (4.44) as follows.

THEOREM–B (4.51)
For $X \in \mathcal{S}_2'$ there is a canonical homotopy class of weak equivalences

$$\mathcal{R} : \Omega C(X) \xrightarrow{\sim} F^*A^*(X) \qquad \textit{in } \langle \mathrm{dgHh}\rangle_1'\,.$$
□

Remark (4.52)
We show later (4.93) that F^* (4.50) is a homotopical equivalence. We use this to give a proof of (4.26).—It is clear that the forgetful functor $\phi : \langle \mathrm{dgHh}\rangle_1' \to \langle \mathrm{dgA}\rangle_1$ has a faithful (fully) derived functor (1.40). Hence the composition F^* :

$\langle \mathrm{cdg^*A}\rangle'_{\langle 2\rangle} \to \langle \mathrm{dgA}\rangle_1$ has the same property. By definition the latter extends to a cofunctor $F^* : \langle \text{flat dg}^*\mathrm{A}\rangle_0 \to \langle \mathrm{dgA}\rangle_1$ which still preserves weak equivalences, cf. (4.43) (a), (4.4) (c). Hence the derived functor of the embedding $\langle \mathrm{cdg^*A}\rangle'_{\langle 2\rangle} \hookrightarrow \langle \mathrm{dg^*A}\rangle_0$ is faithful yielding the assertion.

§ 3 The Quillen Model

In this section K is any $\mathbb{Q}$-algebra.

N° 1 The homotopical category of dg Lie algebras

A *dg Lie algebra* (*dgL*) is a positive chain complex L, endowed with a chain map $[\,,] : L \otimes L \to L$ satisfying the graded anti-commutativity and Jacoby identity, i.e. the laws

$$\begin{aligned} [x,y] &= -(-1)^{|x||y|}[y,x] \\ [x,[y,z]] &= [[x,y],z] + (-1)^{|x||y|}[y,[x,z]] \end{aligned} \qquad (x,y,z \in L).$$

With the evident notion of morphism we obtain the category $\langle \mathrm{dgL}\rangle$.

The full subcategory of dgLs concentrated in degree 0 is denoted by $\langle \mathrm{LA}\rangle$.

We discuss the relationship between dg Lie algebras and (Hopf) algebras. The key result is the PBW–Theorem, which is proved the same way as the complete filtered version in (3.54). This method is due to Quillen [Q] and applies even for $\mathbb{Z}$-graded objects. For a slightly different argument in a more general context, see App. D, N° 3. There we give more details on the constructions that follow.

Given $L \in \langle \mathrm{dgL}\rangle$ let UL be the universal enveloping algebra. There is by definition a map $j : L \to UL$ which is universal for maps from L into the underlying dgL of a dgA. Note that UL is canonically a quotient of $\mathsf{T}L$. The primitive diagonal on the tensor algebra induces a diagonal on UL, so that $UL \in \langle \mathrm{dgH}\rangle$.—This specializes to the symmetric algebra $\mathsf{S}L$ on a chain complex L.

PBW–Theorem (4.53)

For any $L \in \langle \mathrm{dgL}\rangle$ the map

$$e : \mathsf{S}L \xrightarrow{\approx} UL\,, \qquad x_1 \cdot \ldots \cdot x_n \mapsto \frac{1}{n!} \sum_{\sigma \in S_n} \pm\, j(x_{\sigma(1)}) \cdot \ldots \cdot j(x_{\sigma(n)})$$

where $x_i \in L$, is a chain isomorphism compatible with the diagonals. □

The sign $\pm$ is according to the sign rule, i.e. $(-1)^{\varepsilon(\sigma)}$ where $\varepsilon(\sigma)$ is the parity of the permutation $\sigma \in S_n$.

Corollary (4.54)

As a chain complex L is canonically a retract of UL.

As a chain complex UL is canonically a retract of $\mathsf{T}L$. □

For the second assertion note that a section for $\mathsf{T}L \to \mathsf{S}L$ is given by the same formula used in the definition of e.

The corollary has the important consequence that U preserves and reflects H_*-equivalences. For the preservation use the second assertion and Künneth (4.5).

The PBW–Theorem is also the essential tool in the proof of the following Milnor-Moore-Cartier Theorem. For details and generalizations see App. D, N° 3. There is a generalization to $\mathbb{Z}$-graded objects in case K is a field [Q].

MMC–THEOREM (4.55)

The canonical functors $\mathcal{P} : \langle \mathrm{dgH} \rangle_1 \leftrightarrows \langle \mathrm{dgL} \rangle_1 : U$ *are adjoint equivalences of categories.* □

Given a chain complex V there is a chain map $V \to \mathsf{L}V$ which is universal for chain maps from V into some dgL. The obvious map $\mathsf{L}V \to \mathsf{T}V$ is clearly a universal enveloping algebra and is injective by the PBW–Theorem. Hence $\mathsf{L}V$ is the sub-dgL of $\mathsf{T}V$ generated by V.

An object $L \in \langle \mathrm{dgL} \rangle$ is *free* if there exists a graded Lie algebra isomorphism $L \approx \mathsf{L}V$ for some graded free module V. In a similar (and standard) way free dgL-maps are defined, cf. the case of augmented algebras in § 1, N° 1.

THEOREM (4.56)

The category $\langle \mathrm{dgL} \rangle$ *has a canonical homotopy structure: the weak equivalences are the* H_**-equivalences, cofibrant means free, and all objects are fibrant.*

Moreover $\langle \mathrm{dgL} \rangle_1 \subset \langle \mathrm{dgL} \rangle$ *is a homotopical subcategory.*

Proof. In fact $\langle \mathrm{dgL} \rangle$ is a model category where the cofibrations are the free maps and the fibrations are the maps which are surjective in degrees > 0. The verification is analogous to the case of $\langle \mathrm{dgA} \rangle_0$. Here $M \hookrightarrow M \amalg \mathsf{L}(V \oplus dV)$ is a weak equivalence by (4.54), since $UM \hookrightarrow UM \amalg \mathsf{T}(V \oplus dV)$ is a weak equivalence by proof of (4.1). The second assertion is shown as is the second assertion of (4.1). □

The homotopy relation on $\langle \mathrm{dgL} \rangle_1$ can be described by means of a certain natural cylinder object, due to Tanré ([T], [B]). Using this cylinder object Aubry-Lemaire proved the following theorem ([AL]) which explains this homotopy relation in terms of the universal enveloping functor and (derivation) homotopy in $\langle \mathrm{dgA} \rangle_1$. Applications of this theorem are given in the next section.

AUBRY-LEMAIRE THEOREM (4.57)

Let $f, g : L \rightrightarrows M \in \langle \mathrm{dgL} \rangle_1$ *with* L *free. Then* $f \simeq g$ *iff* $Uf \simeq Ug$. □

N° 2 The Quillen functor

Recall that a 2-reduced space or reduced sGp is *rational* if its homotopy groups are uniquely divisible. The simplicial categories we are going to employ are defined over categories with coproducts and having a canonical projective generator. Thus free simplicial objects are defined, see (A.21). By $F_.$ we denote the prolongation of a functor F to the corresponding simplicial categories.

The *Quillen functor* is the composition $N \mathcal{P}_. \widehat{\mathsf{K}}_. G$ in the following diagram of adjoint functors where in each case the upper arrow denotes the left adjoint. (For a discussion in case $\mathsf{K} = \mathbb{Q}$, see Chp. 3, § 3, N° 4). We use the notation $\lambda = N \mathcal{P}_. \widehat{\mathsf{K}}_.$ so that λG is the Quillen functor.

$$(4.58) \qquad \mathcal{S}_2 \underset{\bar{W}}{\overset{G}{\rightleftarrows}} \langle \mathrm{sGp}\rangle_1 \underset{\mathcal{G}.}{\overset{\widehat{\mathsf{K}}.}{\rightleftarrows}} \langle \mathrm{sCHA}\rangle_1 \underset{\mathcal{P}.}{\overset{\widehat{U}.}{\leftrightarrows}} \langle \mathrm{sLA}\rangle_1 \underset{N}{\overset{N^*}{\leftrightarrows}} \langle \mathrm{dgL}\rangle_1$$

The adjoint functors are explained below. The categories are defined — and endowed with a structure (we = weak equivalences, Ob_c = cofibrant objects, Ob_f = fibrant objects) — as shown in the following table.

$\mathcal{S}_2$	:	2-reduced spaces (i.e. simplicial sets), $we = \pi_* \otimes \mathbb{Q}$-equivalences, Ob_c = all objects, Ob_f = rational objects.
$\langle \mathrm{sGp}\rangle_1$	:	reduced simplicial groups, $we = \pi_* \otimes \mathbb{Q}$-equivalences, Ob_c = free objects, Ob_f = rational objects.
$\langle \mathrm{sCHA}\rangle_1$	:	reduced simplicial complete Hopf algebras which here are assumed to be *rigid* and *normal*, $we = \pi_*\mathcal{P}.$-equivalences, Ob_c = free objects, Ob_f = all objects.
$\langle \mathrm{sLA}\rangle_1$	:	reduced simplicial Lie algebras, $we = \pi_*$-equivalences, Ob_c = free objects, Ob_f = all objects.
$\langle \mathrm{dgL}\rangle_1$	:	reduced dg Lie algebras, $we = \pi_*$-equivalences, Ob_c = free objects, Ob_f = all objects.

Although this is not needed here, we mention that the category of rigid normal CHAs is canonically equivalent to that of rigid CLAs or rigid CGps. When K is a field the condition of normality is automatic; see Chp. 3, § 2.

Theorem (4.59)
Endowed with this structure each of these categories is a homotopical category.

Proof. The rational homotopy structure in $\mathcal{S}_2$ is discussed in (A.19). The case of $\langle \mathrm{sGp}\rangle_1$ is similar, see [Q]. The category $\langle \mathrm{CHA}\rangle$ resp. $\langle \mathrm{LA}\rangle$ has colimits, limits, and a (canonical) projective generator P; further $\mathrm{Hom}(P, A)$ is a Kan complex for every simplicial object A over this category; cf. (A.27). Hence $\langle \mathrm{sCHA}\rangle_1$ resp. $\langle \mathrm{sLA}\rangle_1$ has a "standard" homotopy structure (A.25) and this is the one we are considering here. The case of $\langle \mathrm{dgL}\rangle_1$ is discussed in N° 1. □

The pairs of adjoint functors are the following.

$(G, \bar{W})$	:	the loop functor G has a right adjoint, the classifying space functor $\bar{W}$ ([May]).
$(\widehat{\mathsf{K}}., \mathcal{G}.)$ $(\widehat{U}., \mathcal{P}.)$	} :	prolongations of functors defined in Chp. 3.
(N^*, N)	:	the normalization of a sLA is a dgL (App. B), there is a left adjoint N^*, see [Q | p. 221].

Theorem (4.60)

Each of these pairs is an adjoint homotopical pair.
Each of the right adjoints preserves and reflects weak equivalences; so does G.

Proof. It is well-known and easy to see that G preserves free objects (i.e. GX is always free). The case of N^* is similar [Q|p. 221]. For $\widehat{\mathsf{K}}_.$ and $\widehat{U}_.$ this is clear, since these are prolongations of functors preserving the canonical projective generator (3.45) and coproducts.

That the right adjoints preserve fibrant objects is trivial in case of $\mathcal{P}_.$ and N. For $\mathcal{G}_.$ this follows from the simplicial bijection $\mathcal{G}_.A \approx \mathcal{P}_.A$ (3.43). And for $\bar{W}$ we use that $\bar{W}(H)$ is always a Kan complex satisfying $\pi_*(\bar{W}H) = s_{+1}\pi_*(H)$ [May].

It is also clear that the right adjoints preserve and reflect weak equivalences, while G has this property since $\pi_*(GX) = s_{-1}\pi_*(X)$. It remains to show that $\widehat{\mathsf{K}}_.$, $\widehat{U}_.$ and N^* preserve weak equivalences between free objects. We achieve stronger results in (4.83). □

The Quillen functor λG preserves weak equivalences by (4.60). This is also clear by the natural isomorphism $H_*(\lambda GX) = s_{-1}\pi_*(X) \otimes_{\mathbb{Z}} \mathsf{K}$ which is shown in N° 5. In other words,

$$\lambda G : \mathcal{S}_2 \longrightarrow \langle \mathrm{dgL} \rangle_1$$

is fully homotopical. A cofibrant model $\ell : L_X \xrightarrow{\sim} \lambda G(X)$ is said to be a *Quillen model* for $X \in \mathcal{S}_2$.

The principal result of Quillen's theory states that, in case $\mathsf{K} = \mathbb{Q}$, λG is a (rational) homotopical equivalence; see N° 5.

N° 3 Connection to the chain - loop functor

The following theorem is the main result of this section.

Theorem–C (4.61)

For any free $G \in \langle \mathrm{sGp} \rangle_1$ there are canonical H_*-equivalences

$$\boldsymbol{U(\lambda G)} = U\,N(\mathcal{P}_.\widehat{\mathsf{K}}_.G) \underset{\sim}{\xrightarrow{(Ni)'}} N(\widehat{\mathsf{K}}_.G) \underset{\sim}{\xleftarrow{N(c)}} N(\mathsf{K}_.G) = \boldsymbol{C(G)} \qquad \textit{in } \langle \mathrm{dgHh} \rangle_1 .$$

In fact these dgAs have canonical homotopy Hopf diagonals.

We first explain the maps in this chain which are actually defined for any sGp. We recall that the sAA-map $c : \mathsf{K}_.G \to \widehat{\mathsf{K}}_.G$ is the (degreewise) completion of $\mathsf{K}_.G$ for the augmental filtration. Then $\widehat{\mathsf{K}}_.G \in \langle \mathrm{sCAA} \rangle$ and in fact this is a sCHA where the diagonal is induced by the canonical Hopf algebra structure of the group algebra. Then $i : \mathcal{P}_.\widehat{\mathsf{K}}_.G \hookrightarrow \widehat{\mathsf{K}}_.G$ is the inclusion which is a sLA-map into the underlying sLA of $\widehat{\mathsf{K}}_.G$. It follows that Ni is a dgL-map into the underlying dgL of $N(\widehat{\mathsf{K}}_.G) \in \langle \mathrm{dgA} \rangle_0$ by the commutativity of $\mathcal{EM}$. Thus we have $(Ni)'$, the extension to the universal enveloping algebra.

Proposition (4.62)

For any sLA $\mathfrak{g}$ the canonical dgA-map $(Nj)' : UN(\mathfrak{g}) \to N(U_.\,\mathfrak{g})$ is an H_*-equivalence.

Proof. By (4.54) the map in question is a retract of the canonical dgA-map $\mathsf{T}N(\mathfrak{g}) \to N(\mathsf{T}_{.}\,\mathfrak{g})$ which is an H_*-equivalence by Eilenberg-Zilber (B.7). □

LEMMA (4.63)

Let $f : A \to A'$ be a weak equivalence in $\langle\text{sCHA}\rangle_1$ where A, A' are cofibrant. Then f is a π_*-equivalence. Indeed, the restriction $F^k(f)$ is a π_*-equivalence for all $k \geqslant 0$.

Proof. First note that f is a homotopy equivalence in $\langle\text{sCHA}\rangle$, because all objects are fibrant here. This means (A.24) that f is a *simplicial* homotopy equivalence in $\langle\text{sCHA}\rangle$. Recall (A.29) that the prolongation of any functor (between categories with finite sums) preserves simplicial homotopy. Since $F^k : \langle\text{sCHA}\rangle \to \mathcal{S}$ is a prolongation, we conclude that $F^k(f)$ is a simplicial homotopy equivalence (of spaces). Of course this proves the assertion. □

PROPOSITION (4.64)

For any cofibrant object $A \in \langle\text{sCHA}\rangle_1$ the canonical sAA-map $i' : U_{.}\,\mathcal{P}_{.}\,A \to A$ is a π_*-equivalence.

Proof. Choose a π_*-equivalence $g : \mathfrak{m} \xrightarrow{\sim} \mathcal{P}_{.}\,A \in \langle\text{sLA}\rangle_1$ with $\mathfrak{m}$ free, i.e. g is a cofibrant model. Clearly $i' \circ U_{.}\,g = g' : U_{.}\mathfrak{m} \to A$. Note that $U_{.}$ preserves π_*-equivalences: (4.54), Eilenberg-Zilber, Künneth. Hence $U_{.}\,g$ is a π_*-equivalence. It remains to show the same for g' which is the composite of sAA-maps $c : U_{.}\,\mathfrak{m} \to \widehat{U}_{.}\,\mathfrak{m}$ followed by $g'' : \widehat{U}_{.}\,\mathfrak{m} \to A$ (the adjoint of g). Here c is a π_*-equivalence since $U_{.}\,\mathfrak{m}$ is reduced free, see (4.74). And g'' is a *weak equivalence* in $\langle\text{sCHA}\rangle_1$, because g and the adjunction map $\mathfrak{m} \xrightarrow{\sim} \mathcal{P}_{.}\widehat{U}_{.}\,\mathfrak{m} = \widehat{\mathfrak{m}}$ are π_*-equivalences, see (4.83). Now $\widehat{U}_{.}\,\mathfrak{m}$ and A are cofibrant in $\langle\text{sCHA}\rangle_1$, so by (4.63) also g'' is a π_*-equivalence. This completes the proof. □

Here we use essentially the adjoint homotopical equivalence (discussed in N° 5) between $\langle\text{sCHA}\rangle_1$ and $\langle\text{sLA}\rangle_1$ to reduce to showing that for free reduced sLA $\mathfrak{m}$ the completion map $\mathfrak{m} \to \widehat{\mathfrak{m}}$ is a π_*-equivalence. Also i' can be interpreted as a completion map, see (4.67) (c).

Next we give a proof of Theorem–C. Notice however that the proof of $N(c)$ being an H_*-equivalence is postponed to N° 4.

Proof of (4.61). It is easy to check that $(Ni)'$ is the composite of dgA-maps

$$(4.65) \qquad (Ni)' : U\,N(\mathcal{P}_{.}\,\widehat{\mathsf{K}}_{.}G) \xrightarrow{(Nj)'} N(U_{.}\,\mathcal{P}_{.}\,\widehat{\mathsf{K}}_{.}G) \xrightarrow{N(i')} N(\widehat{\mathsf{K}}_{.}G)\,,$$

where $j : \mathcal{P}_{.}\,\widehat{\mathsf{K}}_{.}G \to U_{.}\,\mathcal{P}_{.}\,\widehat{\mathsf{K}}_{.}G$ is the canonical sLA-map. Now $(Nj)'$ is an H_*-equivalence by (4.62). Since $\widehat{\mathsf{K}}_{.}G$ is a *free* sCHA (4.60), $N(i')$ is an H_*-equivalence by (4.64).

Hence $N(i)'$ is an H_*-equivalence, and by the first part of the main result (4.68) of N° 4 also the dgA-map $N(c)$ is an H_*-equivalence.

It remains to show that these maps are compatible with certain homotopy Hopf diagonals. Here the only technical point is the second part of (4.68). Having this it suffices to show that the following diagram in $\langle\text{dgA}\rangle_1$ commutes.

$$
(4.66)\qquad
\begin{array}{ccccccc}
N(\mathsf{K}_{\textstyle .}G) & \xrightarrow{N(\Delta)} & & N(\mathsf{K}_{\textstyle .}G\otimes\mathsf{K}_{\textstyle .}G) & & \xrightarrow[\sim]{\mathcal{AW}} & N(\mathsf{K}_{\textstyle .}G)\otimes N(\mathsf{K}_{\textstyle .}G) \\
\sim\downarrow N(c) & & \swarrow N(c) & & \searrow N(c\otimes c) & & \sim\downarrow N(c)\otimes N(c) \\
N(\widehat{\mathsf{K}}_{\textstyle .}G) & \xrightarrow{N(\widehat{\Delta})} & N(\widehat{\mathsf{K}}_{\textstyle .}G\,\widehat{\otimes}\,\widehat{\mathsf{K}}_{\textstyle .}G) & \xleftarrow[\sim]{N(c)} & N(\widehat{\mathsf{K}}_{\textstyle .}G\otimes\widehat{\mathsf{K}}_{\textstyle .}G) & \xrightarrow[\sim]{\mathcal{AW}} & N(\widehat{\mathsf{K}}_{\textstyle .}G)\otimes N(\widehat{\mathsf{K}}_{\textstyle .}G) \\
\sim\uparrow N(i') & & & & \nearrow N(i'\otimes i') & & \sim\uparrow N(i')\otimes N(i') \\
N(U_{\textstyle .}\mathcal{P}_{\textstyle .}\widehat{\mathsf{K}}_{\textstyle .}G) & \xrightarrow{N(\varphi)} & & N(U_{\textstyle .}\mathcal{P}_{\textstyle .}\widehat{\mathsf{K}}_{\textstyle .}G\otimes U_{\textstyle .}\mathcal{P}_{\textstyle .}\widehat{\mathsf{K}}_{\textstyle .}G) & & \xrightarrow[\sim]{\mathcal{AW}} & N(U_{\textstyle .}\mathcal{P}_{\textstyle .}\widehat{\mathsf{K}}_{\textstyle .}G)\otimes N(U_{\textstyle .}\mathcal{P}_{\textstyle .}\widehat{\mathsf{K}}_{\textstyle .}G) \\
\sim\uparrow (Nj)' & & & & & & \sim\uparrow (Nj)'\otimes(Nj)' \\
UN(\mathcal{P}_{\textstyle .}\widehat{\mathsf{K}}_{\textstyle .}G) & & & \xrightarrow{\qquad\varphi\qquad} & & & UN(\mathcal{P}_{\textstyle .}\widehat{\mathsf{K}}_{\textstyle .}G)\otimes UN(\mathcal{P}_{\textstyle .}\widehat{\mathsf{K}}_{\textstyle .}G)
\end{array}
$$

In fact, this proves that the maps $(Nj)'$, $N(i')$ and $N(c)$ are dgHh-maps where the homotopy diagonals on $UN(\mathcal{P}_{\textstyle .}\widehat{\mathsf{K}}_{\textstyle .}G) = U(\lambda G)$, $N(U_{\textstyle .}\mathcal{P}_{\textstyle .}\widehat{\mathsf{K}}_{\textstyle .}G)$, $N(\widehat{\mathsf{K}}_{\textstyle .}G)$ and $N(\mathsf{K}_{\textstyle .}G) = C(G)$ are *per definitionem* the images in $\mathrm{Ho}\langle\mathrm{dgA}\rangle_1$ of the rows of (4.66). One checks that these dgAs are indeed degreewise flat, cf. (4.69) (a). These homotopy diagonals are *Hopf* since $U(\lambda G)$ is even a dgH (with the standard diagonal φ). Notice that $C(G)$ is a dgHh also by (4.15). We examine consecutively the three horizontal parts of the diagram.

The upper part: Here Δ is induced by the diagonal $G \to G \times G$ and $\widehat{\Delta}$, the diagonal of the sCHA $\widehat{\mathsf{K}}_{\textstyle .}G$, is such that the left quadrangle commutes (even before normalization). The right quadrangle commutes by naturality of $\mathcal{AW}$. The triangle means that $c \circ (c \otimes c)$ is a completion map, see (3.18).

The middle part: Recall that $\mathcal{P}\widehat{\mathsf{K}}G = \{x \in \widehat{\mathsf{K}}G \mid \widehat{\Delta}(x) = 1\widehat{\otimes}x + x\widehat{\otimes}1\,,\ \varepsilon x = 0\}$, thus for $x \in \mathcal{P}\widehat{\mathsf{K}}G$ resp. its image in $U(\mathcal{P}\widehat{\mathsf{K}}G)$, we have $\widehat{\Delta}i'(x) = \widehat{\Delta}(x) = x\widehat{\otimes}1 + 1\widehat{\otimes}x = c(x \otimes 1 + 1 \otimes x) = c(i' \otimes i')\varphi(x)$ since the diagonal φ of $U(\mathcal{P}\widehat{\mathsf{K}}G)$ is by definition primitive on x. Hence the left quadrangle commutes and the rest is clear.

The lower part: Let $a \in N(\mathcal{P}_{\textstyle .}\widehat{\mathsf{K}}_{\textstyle .}G)$ with $|a| = n$ and let $1_n \in U\mathcal{P}\widehat{\mathsf{K}}G_n$ be the unit. We compute $N(\varphi)(Nj)'(a) = N(\varphi j)(a) = \{\varphi j(a)\} = \{j(a) \otimes 1_n + 1_n \otimes j(a)\}$. Here $\{-\}$ denotes the quotient map onto the normalization. On the other hand $((Nj)' \otimes (Nj)')\varphi(a) = \{j(a) \otimes 1 + 1 \otimes j(a)\}$. One checks that $\mathcal{AW}$ carries the former to the latter (the point is that $\mathcal{AW}$ is unital and natural). Hence the rectangle commutes and the proof is complete. □

Remarks (4.67)

(a) When in (4.66) $\mathcal{AW}$ is replaced by $\mathcal{EM}$ (recall this is a dgA-map, right inverse to $\mathcal{AW}$), another *commutative* diagram is obtained. Of course, the homotopy diagonals defined by the rows of this new diagram are the same.

(b) The middle part of the diagram holds with $\widehat{\mathsf{K}}_{\textstyle .}G$ replaced by any cofibrant $A \in \langle\mathrm{sCHA}\rangle_1$. In fact, any such A is homotopy equivalent to some $\widehat{\mathsf{K}}_{\textstyle .}G$ as above, see (4.85). This implies that F^kA is $(k-1)$-connected and consequently that $c : A \otimes A \to A \,\widehat{\otimes}\, A$ is a π_*-equivalence, cf. (4.63) and proof of (4.68). We get a functor $N : \langle\text{free sCHA}\rangle_1 \longrightarrow \langle\mathrm{dgHh}\rangle_1$ and our proof yields the natural weak equivalence

$N(i') : N(U_{\cdot}\mathcal{P}_{\cdot}A) \xrightarrow{\sim} N(A)$ in $\langle\text{dgHh}\rangle_1$. Similar considerations apply to the lower part of the diagram.

(c) One may ask for a more direct proof of (4.64), resp. for an interpretation of the sAA-map $i' : U_{\cdot}\mathcal{P}_{\cdot}A \to A$. This is possible, but requires more "complete algebra", see Chp. 3, § 2, N° 3. Since we assume A normal, we have $A = \widehat{U}_{\cdot}\mathcal{P}_{\cdot}A$, and in fact i' is just the completion map, where *here* $U_{\cdot}\mathcal{P}_{\cdot}A$ gets the filtration induced by $\mathcal{P}_{\cdot}A = L$ (induced by A) — this is not the augmental filtration in general. By PBW etc. this map is a retract of the completion $\mathsf{T}L \to \widehat{\mathsf{T}}L$. By the arguments in the proof of (4.68) below this completion is a π_*-equivalence if the connectivity of F^kL tends to infinity with k. This follows since L is a filtered retract of A by PBW, see part (b) of this remark.

N° 4 The group algebra of a free simplicial group

In this subsection we allow K to be any commutative ring. Recall that for $G \in \langle\text{sGp}\rangle$ the group algebra functor (applied in each degree) yields the sAA $\mathsf{K}_{\cdot}G$. The completion (for the augmental filtration) is the sCAA $\widehat{\mathsf{K}}_{\cdot}G$. Our main concern here is to examine the completion map $c : \mathsf{K}_{\cdot}G \to \widehat{\mathsf{K}}_{\cdot}G$ in case G is free reduced. The main result is the following which is used in N° 3.

THEOREM (4.68)

Let $G \in \langle\text{sGp}\rangle_1$ be free. The completion map $\mathsf{K}_{\cdot}G \to \widehat{\mathsf{K}}_{\cdot}G$ is a π_*-equivalence, and so is the completion map $\widehat{\mathsf{K}}_{\cdot}G \otimes \widehat{\mathsf{K}}_{\cdot}G \to \widehat{\mathsf{K}}_{\cdot}G \mathbin{\widehat{\otimes}} \widehat{\mathsf{K}}_{\cdot}G$.

Proof. Put $A = \mathsf{K}_{\cdot}G$ and regard this as a filtered simplicial module. Recall that a surjective map of simplicial modules is a Kan fibration, cf. (A.28). We show in (4.72) that

(*) the connectivity of F^kA tends to infinity with k.

We claim that in this case the simplicial module map $A \to \widehat{A}$ is a π_*-equivalence.

By a long exact homotopy sequence (LEHS) argument (*) is equivalent to $\pi_q(A) \cong \pi_q(A/F^kA)$ for any fixed q and k large. The isomorphism $A/F^kA = \widehat{A}/F^k\widehat{A}$ shows that the inverse system $\{\pi_q(\widehat{A}/F^k\widehat{A})\}_{k=0,1,\dots}$ becomes eventually constant, so its $\lim^1$-module is zero. This $\lim^1$ term is ([Q|p. 217]) the kernel of the natural surjection $\lim_k\{\pi_q(\widehat{A}/F^k\widehat{A})\} \to \pi_q \lim\{\widehat{A}/F^k\widehat{A}\} = \pi_q(\widehat{A})$. Hence $\pi_q(\widehat{A}/F^k\widehat{A}) = \pi_q(\widehat{A})$ for k large and we conclude that $\pi_q(A) = \pi_q(\widehat{A})$ which proves our claim and the first assertion. Another LEHS argument shows that in fact $F^kA \to F^k\widehat{A}$ is a π_*-equivalence for all k. We conclude that

(**) the connectivity of $F^k\widehat{A}$ tends to infinity with k.

Putting $V = \widehat{A}$ it remains to show that $V \otimes V \to V \mathbin{\widehat{\otimes}} V$ is a π_*-equivalence. We use the simple fact (C.26) that $V \mathbin{\widehat{\otimes}} V = \lim\{V/F^kV \otimes V/F^kV\}_{k=0,1,\dots}$. By Eilenberg - Zilber and Künneth, (**) implies that $\pi_q(V \otimes V) = \pi_q(V/F^kV \otimes V/F^kV)$ for k large. Hence the inverse system $\{\pi_q(V/F^kV \otimes V/F^kV)\}_{k=0,1,\dots}$ becomes eventually constant. As above this implies that, for k large, $\pi_q(V/F^kV \otimes V/F^kV) = \pi_q(V \mathbin{\widehat{\otimes}} V)$. This completes the proof. □

Remark (4.69)

(a) Since we make use of Künneth (4.5) we need that V and V/F^kV are (degreewise) flat where $V = \widehat{\mathsf{K}}_{.}G$. This follows from (3.27) (d) and the fact that any product of copies of K is flat (standard homological algebra).

(b) Since both $A = \mathsf{K}_{.}G$ and $\widehat{A}$ are flat the Künneth Theorem shows that also $c \otimes c : A \otimes A \to \widehat{A} \otimes \widehat{A}$ is a π_*-equivalence; hence so is the completion map $A \otimes A \to \widehat{A} \mathbin{\widehat{\otimes}} \widehat{A}$, cf. (4.66). This map is identified with $\mathsf{K}_{.}(G \times G) \to \widehat{\mathsf{K}}_{.}(G \times G)$.

(c) The second part of the theorem follows immediately from the first of the proof and the previous remark (b) provided $(*)$ is verified for $A = \mathsf{K}_{.}G \otimes \mathsf{K}_{.}G$. This can be done using (C.22).

Definition (4.70)

Let $(X, *)$ be a pointed set. Let FX (resp. VX) be the free group (resp. module) generated by X with the single relation $* = 1$ (resp. $* = 0$). Regard F and V as functors on $\langle\text{Set}\rangle^*$. Then we get the prolongations $F_{.} : \mathcal{S}_0 \to \langle\text{sGp}\rangle$ and $V_{.} : \mathcal{S}_0 \to s\mathcal{M}od$. The former is the well-known *Milnor construction*.

Let $(X, *)$ be a pointed space (= simplicial pointed set). We may regard $(F_{.}X)_n$ as the free group generated by $X_n \smallsetminus *_n$; and $(V_{.}X)_n$ as the free module generated by $\{T_s \mid s \in X_n \smallsetminus *_n\}$.

Clearly $V_{.}X = (F_{.}X)_{\text{ab}} \otimes_{\mathbb{Z}} \mathsf{K}$ and $N(V_{.}X) = \widetilde{C}(X; \mathsf{K})$. We mention $F_{.}X = GSX$ (where S = reduced suspension), cf. [Cu2].

THEOREM (4.71)

For $X \in \mathcal{S}_0$ the graded module map $\theta : V_{.}X \to \overline{\mathsf{K}F_{.}X}$, $T_s \mapsto s - 1$ ($s \in X_n \smallsetminus *_n$), is a natural simplicial map. It induces a natural simplicial isomorphism (of simplicial $\mathsf{K}F_{.}X$-modules)

$$\Theta^k : \mathsf{K}F_{.}X \otimes (V_{.}X)^{\otimes k} \xrightarrow{\simeq} \overline{\mathsf{K}F_{.}X}^{\,k}.$$

Proof. By (3.27) it suffices to show that $\theta : VX \to \overline{\mathsf{K}FX}$ is natural as X runs over $\langle\text{Set}\rangle^*$. This is readily checked. □

In the same way the isomorphisms in (3.27) (b)–(d) yield natural simplicial isomorphisms, but this is not needed.

It remains to verify the connectivity assertion $(*)$ for $A = \mathsf{K}_{.}G$.

THEOREM (4.72)

Let G be a connected free sGp. Then $\overline{\mathsf{K}_{.}G}^{\,k}$ is $(k-1)$-connected.

Proof. Regard $G \mapsto \overline{\mathsf{K}G}^{\,k} =: \Phi(G)$ as a functor from groups to abelian groups. One checks that $\Phi(e) = 0$ and that Φ preserves filtered colimits. By Curtis' connectivity theorem (A.30), when $\Phi_{.}\,F_{.}X$ is $(k-1)$-connected in case X is a finite wedge of 1-spheres $\Delta[1]/(0) \cup (1)$, then $\Phi_{.}\,G$ is $(m+k-1)$-connected for every m-connected free sGp G.—It is therefore more than sufficient to prove the assertion in case $G = F_{.}X$, where X is a reduced space. By (4.71) and Eilenberg-Zilber there is an H_*-equivalence of chain complexes

$$(4.73)\qquad \begin{aligned} N(\overline{\mathsf{K}F_{.}X}^{k}) &\cong N(\mathsf{K}F_{.}X \otimes (V_{.}X)^{\otimes k}) \xrightarrow{\sim} \\ N(\mathsf{K}F_{.}X)\otimes N(V_{.}X)^{\otimes k} &= C(F_{.}X)\otimes \widetilde{C}(X)^{\otimes k}. \end{aligned}$$

Since X is reduced, the latter is obviously k-reduced. Hence passing to homology yields $\pi_*(\overline{\mathsf{K}F_{.}X}^{k}) = 0$ in degrees $< k$ as desired. □

Theorem (4.74)

Let $A \in \langle \mathrm{sAA}\rangle_1$ be free. Then $c : A \to \widehat{A}$ is a π_*-equivalence.

Proof. This is shown by the same method as (4.68). It is sufficient that the connectivity of $\overline{A}^{k}$ tends to infinity with k. By Curtis' connectivity theorem the latter reduces to the (trivial) case $A = \mathsf{T}\,V_{.}X$. □

As an application we give the following simplicial version of the Bott-Samelson Theorem. It follows, by naturality of completion, from (4.68), (4.74) and (4.71) (d).

Corollary (4.75)

Let X be a reduced space, and $F_{.}X$, $V_{.}X$ and θ as in (4.71). Then θ induces a π_*-equivalence $\mathsf{T}\,V_{.}X \xrightarrow{\simeq} \mathsf{K}F_{.}X$ of reduced sAAs. Hence $H_*(F_{.}X) = \mathsf{T}\overline{H}_*(X)$ if $H_*(X)$ is flat. □

The cobar spectral sequence (4.76)

Recall that $G = GX$ has (in each degree) a canonical freely generating subset. Hence the natural graded module map $\theta : G_{\mathrm{ab}} \otimes \mathsf{K} \to \overline{\mathsf{K}_{.}G}$ can be defined as in (3.27), but this is *not* a simplicial map (because of d_0). The composition with $\overline{\mathsf{K}_{.}G} \twoheadrightarrow \mathrm{gr}^1\mathsf{K}_{.}G$ however *is* simplicial, as is shown by direct computation. By (3.27) (c) there results a *simplicial* graded algebra isomorphism

$$\mathsf{T}_{.}(G_{\mathrm{ab}}\otimes \mathsf{K}) \xrightarrow{\cong} \mathrm{gr}\,\mathsf{K}_{.}G$$

where $\mathsf{K}_{.}G$ is filtered by the augmental filtration.

Applying normalization we get a multiplicative spectral sequence E with

$$E^1_{-p,q} = H_q(\widetilde{C}(X)^{\otimes p}).$$

Here we use $N(GX)_{\mathrm{ab}} \otimes \mathsf{K} = s_{-1}\widetilde{C}(X)$ which is well-known and easily verified. When X is 2-reduced, we have finite convergence to $H_*(GX;\mathsf{K})$ since in this case (4.72) implies that the filtration of $\mathsf{K}_{.}G$ is homologically finite ([HiSt | Thm. 3.5]).

These and further properties of the spectral sequence $E = E(X)$ are stated without proof in [Cu2 | Sec. 10] where it is called the *cobar spectral sequence*. In fact, E is canonically isomorphic (from E^1 onwards) to the spectral sequence derived from the augmental filtration of $\Omega C(X)$. This is proved as follows. — It is clear that the natural map $\mathcal{B} : \Omega C(X) \to C(GX)$ of (4.10) is filtration preserving ($X \in \mathcal{S}_1$). By a standard spectral sequence argument and Künneth it suffices to show that $E^0_{-1,*}(\mathcal{B})$ is chain homotopic to $-\mathrm{id}$, when regarded as a self-map of $\widetilde{C}(X)$. By construction this self-map is natural and equals $-\mathrm{id}$ in degree 1, hence an acyclic model argument yields the result.

This proves anew that $\mathcal{B}$ is an H_*-equivalence for $X \in \mathcal{S}_2$.

N° 5 A proof of the Quillen equivalence

Using our result on the group algebra of a free sGp we present a simplified proof of the Quillen equivalence and related results. Some of these results are used in the proof (4.60). In particular we show that $H_*(\lambda GX) = s_{-1}\pi_*(X) \otimes_{\mathbb{Z}} \mathsf{K}$ as graded Lie algebras.

Let G be a reduced free sGp. Regard the following diagram of simplicial maps.

(4.77)
$$\begin{array}{ccccc} G & \xrightarrow{u_G} & \mathcal{G}.\widehat{\mathsf{K}}.G & \xrightarrow[\approx]{\log} & \mathcal{P}.\widehat{\mathsf{K}}.G \\ h\downarrow & & j\downarrow & \triangle & \nearrow i \\ \mathsf{K}.G & \xrightarrow{c} & \widehat{\mathsf{K}}.G & & \end{array}$$

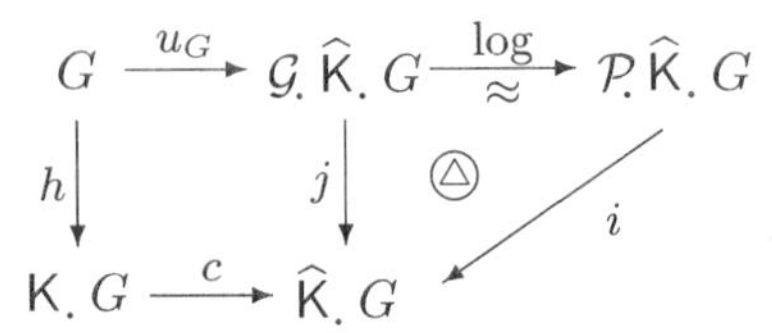

Here u_G is the adjunction map which is the restriction of c to the grouplike elements. Moreover h and j are simplicial maps defined by $x \mapsto x - 1$. The left square in (4.77) is thus commutative.

Recall that if Λ is a simplicial (Lie) algebra, then $\pi_*(\Lambda) = H_*(N\Lambda)$ is a graded (Lie) algebra. In a similar way (i.e. using shuffles) $\pi_*(G)$ is a graded Lie algebra in a natural way, via the *Samelson product* [Cu2|p. 197]. In case $G = GX$ this graded Lie algebra structure on $\pi_*(GX) = s_{-1}\pi_*(X)$ corresponds to the Whitehead product on the homotopy groups of X.

LEMMA (4.78)

The sub-diagram $\triangle$ becomes commutative when π_* is applied.
The gL-map $\pi_*(i) = i_*$ is injective.

Proof. Here one may replace $\mathsf{K}.G = A$ by any reduced sCHA. We quote the proof from [Q|p. 225]. The map i_* is injective since there is a natural retraction of a CHA onto its primitive CLA. For another argument see proof of (4.79). For the first assertion we have to prove the following claim: given $x \in N(\mathcal{P}.A)_k$ where $k > 0$, then x and $e^x - 1$ differ by a boundary in NA. (Here we regard the normalization as a sub-complex). Let $S^k = \Delta[k]/\dot{\Delta}[k]$ and σ its canonical k-simplex. There is a unique sCHA-map $\widehat{\mathsf{T}}.V.S^k \to A$ sending σ to x. By naturality of exp it suffices to prove the claim for σ which indeed is an element of $N(\mathcal{P}.\widehat{\mathsf{T}}.V.S^k)$. But by (4.74) $\pi_*(\widehat{\mathsf{T}}.V.S^k) = \pi_*(\mathsf{T}.V.S^k)$ and by Künneth this is a tensor algebra on the class of σ. In particular if I is the (degreewise) augmentation ideal of $\mathsf{T}.V.S^k$ then $\pi_k(I^2) = 0$. Therefore $(e^\sigma - 1) - \sigma \in N(I^2)_k$ is a boundary. □

When $f : B \to V$ is a map from an abelian group to a K-module, there is a canonical map $B \otimes_{\mathbb{Z}} \mathsf{K} \to V$ which we also denote by f.

THEOREM (4.79)

$\ell = (\log \circ u_G)_* : \pi_*(G) \otimes_{\mathbb{Z}} \mathsf{K} \to \pi_*(\mathcal{P}.\widehat{\mathsf{K}}.G) = H_*(\lambda G)$ *is a gL-isomorphism.*

Proof. Notice that $\pi_*(\mathsf{K}.G)$ resp. $\pi_*(\widehat{\mathsf{K}}.G)$ is a gL via commutator. It is clear that i_* and c_* are gL-maps. Further $h_* : \pi_*(G) \otimes_{\mathbb{Z}} \mathsf{K} \to \pi_*(\mathsf{K}.G) = H_*(G;\mathsf{K})$ is the Hurewicz map which is well-known to be a gL-map (in the simplicial setting this is immediate from the definitions). Hence the lemma shows that ℓ is a gL-map. Now we have the following commutative diagram where h_*' etc. denotes the extension to the universal enveloping algebra.

$$(4.80)\qquad \begin{array}{ccc} U(\pi_*(G)\otimes_{\mathbb{Z}}\mathsf{K}) & \xrightarrow{U(\ell)} & UH_*(\lambda G) \\ {\scriptstyle h_*'}\downarrow{\scriptstyle \simeq} & & {\scriptstyle \simeq}\downarrow{\scriptstyle i_*'} \\ H_*(G;\mathsf{K}) & \underset{c_*}{\xrightarrow{\simeq}} & \pi_*(\widehat{\mathsf{K}}_{.}\,G) \end{array}$$

Recall that c_* is an isomorphism by (4.68). Observe that $i_*{}'$ is the composite of $UH_*(\lambda G) = H_*(U\lambda G)$ followed by the isomorphism $H_*((Ni)')$ of (4.61). Hence $i_*{}'$ is an isomorphism. Finally $h_*{}'$ is an isomorphism by the Milnor - Moore Theorem (see remark). □

Remark (4.81)

We use here the Milnor - Moore (Cartan - Serre) Theorem. In the topological setting it states ([MiM]) that $U(\pi_*(G)\otimes_{\mathbb{Z}}\mathsf{K}) = H_*(G;\mathsf{K})$ for any connected (topological) H-space G. The general case follows from the rational by the Universal Coefficient Theorem. The simplicial version follows by geometric realization ([Q | p. 224ff]).

We consider the adjunction maps

$$(4.82)\qquad \begin{array}{lll} u_X : X \to \bar{W}GX & , & X\in\mathcal{S}_2 \\ u_G : G \to \mathcal{G}_{.}\,\widehat{\mathsf{K}}_{.}\,G & , & G\in\langle\mathrm{sGp}\rangle_1 \ \text{ free} \\ u_{\mathfrak{m}} : \mathfrak{m} \to \mathcal{P}_{.}\,\widehat{U}_{.}\,\mathfrak{m} & , & \mathfrak{m}\in\langle\mathrm{sLA}\rangle_1 \ \text{ free} \\ u_L : L \to N\,N^*L & , & L\in\langle\mathrm{dgL}\rangle_1 \ \text{ free.} \end{array}$$

Theorem (4.83)

The maps u_X, $u_{\mathfrak{m}}$, u_L are weak equivalences, and so is u_G in case $\mathsf{K}=\mathbb{Q}$. In general

$$(*)\qquad (u_G)_* : \pi_*(G)\otimes_{\mathbb{Z}}\mathsf{K} \xrightarrow{\simeq} \pi_*(\mathcal{G}_{.}\,\widehat{\mathsf{K}}_{.}\,G)$$

where $\pi_*(\mathcal{G}_{.}\,\widehat{\mathsf{K}}_{.}\,G)$ is a K-module via $\log_* : \pi_*(\mathcal{G}_{.}\,\widehat{\mathsf{K}}_{.}\,G)\cong\pi_*(\mathcal{P}_{.}\,\widehat{\mathsf{K}}_{.}\,G)$.

Recall that the chain (4.58) of adjoint pairs from $\mathcal{S}_2$ to $\langle\mathrm{dgL}\rangle_1$ is one of adjoint homotopical pairs. Hence by the Adjoint Functor Theorem we obtain a chain of derived adjoint pairs

$$(4.84)\qquad \mathrm{Ho}\,\mathcal{S}_2 \underset{\bar{W}^{\sim}}{\overset{\widetilde{G}}{\rightleftarrows}} \mathrm{Ho}\langle\mathrm{sGp}\rangle_1 \underset{\widetilde{\mathcal{G}_{.}}}{\overset{(\widehat{\mathsf{K}}_{.})^{\mathsf{L}}}{\rightleftarrows}} \mathrm{Ho}\langle\mathrm{sCHA}\rangle_1 \underset{\widetilde{\mathcal{P}_{.}}}{\overset{(\widehat{U}_{.})^{\mathsf{L}}}{\leftrightarrows}} \mathrm{Ho}\langle\mathrm{sLA}\rangle_1 \underset{\widetilde{N}}{\overset{(N^*)^{\mathsf{L}}}{\leftrightarrows}} \mathrm{Ho}\langle\mathrm{dgL}\rangle_1\,.$$

As GX is always cofibrant, the composite from left to right is the fully derived of the Quillen functor,

$$\widetilde{\lambda G} = \widetilde{N}\,\widetilde{\mathcal{P}_{.}}\,(\widehat{\mathsf{K}}_{.})^{\mathsf{L}}\,\widetilde{G}\,.$$

By (4.83) and the Adjoint Functor Theorem we see that (4.84) is a chain of adjoint equivalences of categories, where for $(\widehat{\mathsf{K}}_{.}^{\mathsf{L}},\widetilde{\mathcal{G}_{.}})$ we have to assume $\mathsf{K}=\mathbb{Q}$. In different terms (1.31) this reads as follows:

QUILLEN EQUIVALENCE (4.85)

Each of the pairs $(G, \bar{W})$, $(\widehat{U}_{.}, \mathcal{P}_{.})$ *and* (N^*, N) *consists of adjoint homotopical equivalences. The same holds for* $(\widehat{\mathsf{K}}_{.}, \mathcal{G}_{.})$ *in case* $\mathsf{K} = \mathbb{Q}$.

Hence λG *is a homotopical equivalence in case* $\mathsf{K} = \mathbb{Q}$. □

Proof of (4.83) *in outline.* It is a classical [K4] that u_X is even a π_*-equivalence.

The assertion $(*)$ about u_G follows at once from (4.79).

$\pi_*(\mathfrak{m}) \cong \pi_*(\mathcal{P}_{.}\widehat{U}_{.}\mathfrak{m})$ if $\mathfrak{m}$ is reduced free.—We know that $\mathcal{P}\widehat{U}\mathfrak{m} = \widehat{\mathfrak{m}}$ (3.90). Hence as in the proof of (4.68) one is reduced to show that $\Gamma^k\mathfrak{m}$ is $(k-1)$-connected where $\mathfrak{m} = \mathsf{L}V_{.}X$ with $X \in \mathcal{S}_1$. But if V is any K-module, $\mathsf{L}V$ is naturally a retract of $\mathsf{T}V$ in such a way that $\Gamma^k\mathsf{L}V$ is a retract of $\overline{\mathsf{T}V}^k$. Such a retraction is given by

$$v_1 \otimes \ldots \otimes v_n \mapsto \frac{1}{n}\left[v_1, [v_2, \ldots [v_{n-1}, v_n] \ldots]\right],$$

cf. (3.53). Now $\overline{\mathsf{T}V_{.}X}^k$ is $(k-1)$-connected by proof of (4.74) yielding the assertion.

$H_*(L) \cong H_*(NN^*L)$ if L is reduced free.—Here we only remark that this statement and the proof given in [Q|I.4] have an analog for dgAs, but the proof is easier and works over any commutative ring; from this the assertion is easily deduced (cf. [ShT|§4]). □

Remarks (4.86)

In case $\mathsf{K} = \mathbb{Q}$ Quillen proves $(*)$ by the same method as (4.68) using the difficult connectivity result for $\Gamma^k G$ ([Cu], [Q3]). Since $\mathcal{G}\widehat{\mathbb{Q}}G = \widehat{G}$ is the $\mathbb{Q}$-completion of G (3.90) this result of Quillen's is also contained in [BouK] where the method is essentially the same. If $(*)$ is proved this way, we obtain the Milnor - Moore Theorem (4.81) as a consequence of our results, cf. (4.80).

§4 MAIN RESULTS

Let K be a field of characteristic 0, or at least a hereditary $\mathbb{Q}$-algebra.

N°1 Summary

Recall that $\lambda G : \mathcal{S}_2 \to \langle \mathrm{dgL} \rangle_1$ is the Quillen functor, $A^* : \mathcal{S}_2' \to \langle \mathrm{cdg^*A} \rangle'_{\langle 2 \rangle}$ the Sullivan cofunctor, and $\Omega C : \mathcal{S}_2 \to \langle \mathrm{dgHh} \rangle_1$ the cobar - chain functor. Composition with $U : \langle \mathrm{dgL} \rangle_1 \to \langle \mathrm{dgHh} \rangle_1$ resp. $F^* : \langle \mathrm{cdg^*A} \rangle'_{\langle 2 \rangle} \to \langle \mathrm{dgHh} \rangle'_1$ yields three functors to the category $\langle \mathrm{dgHh} \rangle_1$ of Hopf algebras up to homotopy, cf. diagram (4.88). Recall that $\langle \mathrm{dgHh} \rangle'_1$ is the full subcategory of $\langle \mathrm{dgHh} \rangle_1$ whose objects have homology of finite type. The loop cofunctor F^* is essentially the (functional) dual of the bar construction. We regard $\mathcal{S}_2$ as a homotopical category with its rational structure, and $\mathcal{S}_2'$ as a homotopical subcategory.

We summarize our main theorems of §1–§3. First we recall Theorem–B (4.51).

THEOREM–B

For $X \in \mathcal{S}_2'$ *there is a canonical homotopy class of* H_**-equivalences*

$$\mathcal{R} : \Omega C(X) \xrightarrow{\sim} F^* A^*(X) \qquad in\ \langle \mathrm{dgHh} \rangle'_1 .$$

□

Combining Theorem–A (4.18) with Theorem–C (4.61) yields ...

Theorem–AC

For $X \in \mathcal{S}_2$ there is a canonical homotopy class of H_*-equivalences

$$\Gamma_X : \Omega C(X) \xrightarrow{\sim} U(\lambda G(X)) \qquad \textit{in } \langle \mathrm{dgHh} \rangle_1 .$$

□

In particular Γ is natural up to homotopy: given $f : X \to Y$ there is a (derivation) homotopy $U\lambda G(f) \circ \Gamma_X \simeq \Gamma_Y \circ \Omega C(f)$. Indeed there is unique (up to homotopy) Γ_X satisfying $\Gamma_X \circ (Ni)' \simeq N(c) \circ \mathcal{B}$ where these maps are taken from Theorem–A and C. We use here that $\Omega C(X)$ is cofibrant. Note that Theorem–AC holds over any $\mathbb{Q}$-algebra.

The following is essentially a reformulation of these theorems.

Theorem (4.87)

The following diagram of fully homotopical functors commutes up to natural weak equivalence. The homotopy classes of these weak equivalences are canonical.

Hence the associated diagram of fully derived functors commutes up to canonical isomorphism.

□

(4.88)

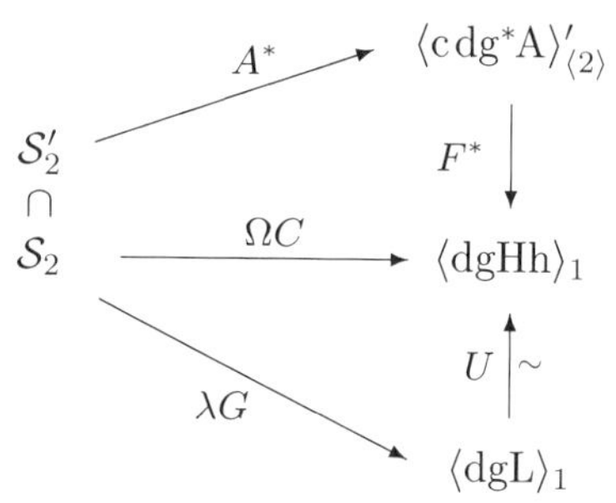

The significance of this result is increased considerably by the fact that here *U is a homotopical equivalence* (Anick's equivalence, see N° 2).

Hence ΩC yields an essentially well-defined functor from $\mathrm{Ho}\,\mathcal{S}_2$ into $\mathrm{Ho}\langle \mathrm{dgL} \rangle_1$, and analogously for A^*. We conclude that *these functors are canonically isomorphic to the derived of λG*. This proves and extends the Baues - Lemaire conjecture, see N° 3.

Since the Quillen functor is a homotopical equivalence, Anick's equivalence and (4.87) show that also *ΩC is a homotopical equivalence*. We also easily deduce that *A^* is a homotopical equivalence*, see N° 4.

N° 2 Anick's equivalence

Recall that $\mathbf{dgHh}_1 \subset \langle \mathrm{dgHh} \rangle_1$ resp. $\mathbf{dgL}_1 \subset \langle \mathrm{dgL} \rangle_1$ denote the homotopical subcategories of cofibrant (- fibrant) objects. Recall that these category embeddings are homotopical equivalences. Hence the equivalence indicated in diagram (4.88) is equivalent to ...

ANICK'S EQUIVALENCE

The universal enveloping functor is a homotopical equivalence

$$U : \mathbf{dgL}_1 \xrightarrow{\sim} \mathbf{dgHh}_1 .$$

□

This in turn follows immediately from the two theorems below which give some more. For a sketch of proof and generalizations, see App. D, N° 4.

ANICK'S FIRST THEOREM

Let $(A, \Psi) \in \mathbf{dgHh}_1$. *Then there is a Hopf diagonal* χ *on* A *with* $[\chi] = \Psi$. □

Note that $(A, \chi) = U(L)$ with $L \in \mathbf{dgL}_1$ by the MMC–Theorem (4.55).

ANICK'S SECOND THEOREM

Let L, $M \in \langle \mathrm{dgL} \rangle_1$ *and let* L *be free. Let* $f : UL \to UM$ *be a dgHh-map. Then* $f \simeq U(\ell)$ *for some dgL-map* $\ell : L \to M$ *which is unique up to homotopy.* □

The uniqueness (up to homotopy) of f is a consequence of the Aubry - Lemaire theorem (4.57).

N° 3 Proof of the Baues - Lemaire conjecture

We essentially reformulate (4.87) in terms of (cofibrant) models and combine it with Anick's equivalence.

Recall that $\mathbf{M}'_2$ is the category of 2-reduced KS-algebras of finite type. We claim that the following diagram commutes where the horizontal maps are the embeddings of the respective subcategories of cofibrant objects. (The functor $\mathcal{L}$ is defined below.)

$$\text{(4.89)} \qquad \begin{array}{ccc} \langle \mathrm{cdg^*A} \rangle'_{\langle 2 \rangle} & \overset{\sim}{\hookleftarrow} & \mathbf{M}'_2 \\ {\scriptstyle F^*} \downarrow & & \downarrow {\scriptstyle \Omega\#} \\ \langle \mathrm{dgHh} \rangle_1 & \overset{\sim}{\hookleftarrow} & \mathbf{dgHh}_1 \\ {\scriptstyle U} \uparrow {\scriptstyle \sim} & & {\scriptstyle U} \uparrow {\scriptstyle \sim} \\ \langle \mathrm{dgL} \rangle_1 & \overset{\sim}{\hookleftarrow} & \mathbf{dgL}_1 \end{array} \qquad \mathbf{M}'_2 \xrightarrow{\mathcal{L}\#} \mathbf{dgL}_1$$

This is seen as follows. Recall that $\Omega\#$ is the composition of the (functional) dualization cofunctor $\# : \mathbf{M}'_2 \to \langle \mathrm{cdgC} \rangle_2$ with the cobar construction $\Omega : \langle \mathrm{cdgC} \rangle_2 \to \mathbf{dgHh}_1$. Recall that $\Omega(C)$ has a canonical Hopf diagonal if C is cocommutative. As usual let $\mathcal{L} = \mathcal{P}\Omega$ where $\mathcal{P} : \langle \mathrm{dgH} \rangle_1 \to \langle \mathrm{dgL} \rangle_1$. Then $U\mathcal{L} = \Omega$ by the MMC–Theorem. Moreover, the cofunctor F^* extends $\Omega\#$, see (4.47).

Remark. It is clear that all the functors in (4.89) preserve weak equivalences, since we know this for F^* and U. Hence there is an induced diagram of fully derived functors. Note that the derived category of $\mathbf{M}'_2$ is just the quotient category $\mathbf{M}'_2/\simeq$, etc.

THEOREM (BAUES - LEMAIRE CONJECTURE) (4.90)

For $X \in \mathcal{S}_2'$ choose a 2-reduced free model of finite type $\ell_X : M_X \xrightarrow{\sim} A^*(X)$; and a free model $\mu_X : L_X \xrightarrow{\sim} \lambda G(X)$. Then there is a canonical homotopy class of H_*-equivalences

$$\alpha_X : L_X \xrightarrow{\sim} \mathcal{L}\#(M_X) \qquad \textit{in } \langle \mathrm{dgL} \rangle_1 .$$

In particular we have the naturality (up to homotopy) of α. This is actually not part of the conjecture in [BL]. Since $\mathcal{L}\#(M_X)$ and L_X are free, α_X is a homotopy equivalence. The theorem also extends the statement of the conjecture, in that it is proved over any hereditary $\mathbb{Q}$-algebra.

Proof. First we construct a weak equivalence from $U(L_X)$ to $U(\mathcal{L}\#(M_X)) = F^*(M_X)$ in $\langle \mathrm{dgHh} \rangle_1$. By virtue of Theorem–B and Theorem–AC, and since U and F^* preserve weak equivalences, we obtain a chain of weak equivalences

$$U(L_X) \xrightarrow{\sim} U(\lambda G X) \xleftarrow{\sim} \Omega C(X) \xrightarrow{\sim} F^* A^*(X) \xrightarrow{\sim} F^*(M_X) \quad \text{in } \langle \mathrm{dgHh} \rangle_1 .$$

and all these are natural up to homotopy. Since $U(L_X)$ is cofibrant, we obtain a weak equivalence, f_X say, representing the isomorphism in $\mathrm{Ho}\langle \mathrm{dgHh} \rangle_1$ induced by that chain. By Anick's second theorem $f_X \simeq U(\alpha_X)$ for some α_X which is the desired weak equivalence in $\langle \mathrm{dgL} \rangle_1$. □

We now employ the *cellular Lie algebra model* (App. D). Note that Anick's first theorem implies that $\Omega C(X) = U(\mathcal{M}_X)$ for some free dgL $\mathcal{M}_X$. In App. D we show that the construction of $\mathcal{M}_X$ can be done in a *natural* way yielding the functor $\mathcal{M} : \mathcal{S}_2 \to \mathbf{dgL}_1$ satisfying $U\mathcal{M} = \Omega C$. Then Theorem–B, Theorem–AC and Anick's equivalence yield the following two results, which together imply the Baues - Lemaire conjecture.

THEOREM (4.91)

For $X \in \mathcal{S}_2'$ choose a 2-reduced model of finite type $\ell_X : M_X \xrightarrow{\sim} A^*(X)$. Then there is a canonical homotopy class of H_*-equivalences

$$\rho_X : \mathcal{M}(X) \xrightarrow{\sim} \mathcal{L}\#(M_X) \qquad \textit{in } \langle \mathrm{dgL} \rangle_1 .$$

□

By definition, $U(\rho_X) \simeq F^*(\ell_X) \circ \mathcal{R}_X$. Up to homotopy, any family of maps ρ thus obtained is natural, and independent of the choice of $\mathcal{R}$.

THEOREM (4.92)

For $X \in \mathcal{S}_2$ there is a canonical homotopy class of H_*-equivalences

$$\gamma_X : \mathcal{M}(X) \xrightarrow{\sim} \lambda G(X) \qquad \textit{in } \langle \mathrm{dgL} \rangle_1 .$$

□

By definition, $U(\gamma_X) \simeq \Gamma_X$. Up to homotopy, any family of maps γ thus obtained is natural, and independent of the choice of Γ.

N° 4 Rational equivalence

Let $\mathbf{dgL}_1'$ be the homotopical subcategory of $\mathbf{dgL}_1$ consisting of those objects having homology of finite type. Analogously define $\mathbf{dgHh}_1'$.

THEOREM (4.93)
The following are homotopical equivalences:

$$\begin{aligned} \mathcal{L}\# &: \mathbf{M}_2' \xrightarrow{\sim} \mathbf{dgL}_1' \\ \Omega\# &: \mathbf{M}_2' \xrightarrow{\sim} \mathbf{dgHh}_1' \\ F^* &: \langle \mathrm{cdg^*A} \rangle_2' \xrightarrow{\sim} \langle \mathrm{dgHh} \rangle_1' \end{aligned}$$

Proof. The first assertion is well-known ([T], [Q | App. B]). The second and third follows by Anick's equivalence, see diagram (4.89). □

Now let $\mathsf{K} = \mathbb{Q}$. Consider once more diagram (4.88) and be aware of (4.93). From Quillen's result that λG is a homotopical equivalence (§ 3, N° 5) we can now deduce the same for ΩC and A^*. The latter is the well-known Sullivan equivalence. The former seems new and reads as follows. (There is a similar result in the topological context, see [An 2]).

THEOREM (4.94)
The cellular algebra model $\Omega C : \mathcal{S}_2 \to \mathbf{dgHh}_1$ *is a homotopical equivalence.*
Hence:
The cellular Lie algebra model $\mathcal{M} : \mathcal{S}_2 \to \mathbf{dgL}_1$ *is a homotopical equivalence.*
□

Appendix D THE CELLULAR LIE ALGEBRA MODEL

Here K is an arbitrary commutative ring.

Given $A \in \langle \mathrm{dgA} \rangle$ or $\langle \mathrm{dgL} \rangle$ or $\langle \mathrm{dgH} \rangle$, we use $A_{(m)}$ to denote the sub-object of A generated by all elements of degree *strictly* less than m.

In N° 1 we construct a *natural* diagonal ψ for $\Omega C(X)$, such that $[\psi] = \Psi$ is the canonical homotopy diagonal. We show that up to *natural* derivation homotopy ψ is coassociative and cocommutative. In N° 2 we apply a natural version of Anick's first theorem and obtain a *natural* diagonal $\chi \simeq \psi$ which is Hopf in a certain 'mild' range of degrees. Hence in case $\mathsf{K} \supset \mathbb{Q}$ we have that $(\Omega C(X), \chi)$ is a Hopf algebra and taking the primitives yields a functor $X \mapsto \mathcal{M}(X)$ into the category of dg Lie algebras.

N° 1 A natural diagonal for the cobar - chain functor

Recall that for $(X, Y) \in \mathcal{S}_0 \times \mathcal{S}_0$ there is a natural EZ-morphism (B.10) in $\langle \mathrm{dgC} \rangle_0$

$$^{\xi}\, C(X \times Y) \underset{\mathcal{AW} = \varphi}{\overset{\mathcal{EM} = \nabla}{\rightleftarrows}} CX \otimes CY$$

which we call the *standard* EZ-morphism.

A natural diagonal on $\Omega C(X)$ (D.1)
Recall that for $X \in \mathcal{S}_2$ the (solid) chain of dgA-maps

$$\Omega C(X) \xrightarrow{\Omega C(\Delta)} \Omega C(X \times X) \underset{\sim}{\xleftarrow{\Omega(\nabla)}} \Omega(CX \otimes CX) \xrightarrow{j} \Omega C(X) \otimes \Omega C(X)$$
$$\text{(dashed arrow } F: \Omega C(X\times X) \dashrightarrow \Omega(CX \otimes CX))$$

gives $\Omega C(X)$ a dgHh-structure: the homotopy diagonal is

$$\Psi = [j] \circ [\Omega(\nabla)]^{-1} \circ [\Omega C(\Delta)],$$

see (4.16). By (2.68), $\Omega(\nabla)$ has a natural dgA-retraction F, the *extension* of φ. We define a natural diagonal ψ on $\Omega C(X)$ by

$$\psi = j \circ F \circ \Omega C(\Delta).$$

Then clearly $\Psi = [\psi]$.

THEOREM (D.2)
Let $X \in \mathcal{S}_2$. There are natural derivation homotopies $\Gamma^1 : \psi \simeq \tau\,\psi$ and $\Gamma^2 : (\psi \otimes 1)\psi \simeq (1 \otimes \psi)\psi$. Further ψ is counital, i.e. $\pi_1\,\psi = 1 = \pi_2\,\psi$.

Proof. We use the results of Chp. 2, § 4. The standard EZ-morphism with $X = Y$ yields by (2.68) a natural derivation homotopy $\mathcal{H} : 1_{\Omega C(X \times Y)} \simeq \Omega(\nabla) \circ F$.

Claim:

$$\Gamma^1 = j \circ F \circ \Omega C(\tau) \circ \mathcal{H} \circ \Omega C(\Delta)$$

is the first of the desired homotopies (where $\tau : X \times Y \to Y \times X$ is the interchange map). Indeed, Γ^1 is a derivation homotopy from $j \circ F \circ \Omega C(\tau) \circ \Omega C(\Delta) = j \circ F \circ \Omega C(\Delta) = \psi$ to

$$\begin{aligned} & j \circ F \circ \Omega C(\tau) \circ \Omega(\nabla) \circ F \circ \Omega C(\Delta) \\ = \;& j \circ \{F \circ \Omega(\nabla)\} \circ \Omega(\tau) \circ F \circ \Omega C(\Delta) \\ = \;& \tau \circ j \circ F \circ \Omega C(\Delta) = \tau \circ \psi . \end{aligned}$$

Here we use the commutativity of ∇ and j.

In order to construct Γ^2, regard the following diagram consisting of four natural EZ-morphisms, resulting in the obvious way from instances of the standard EZ-morphism. For example, on the top left one sees the standard EZ-morphism for the pair $(X, Y \times Z)$. The two on the left-hand side and the two on the right-hand side are composable. Each circled capital letter denotes the extension to $\langle \mathrm{dgA} \rangle_1$ of the map nearby, see (2.68).

(D.3)

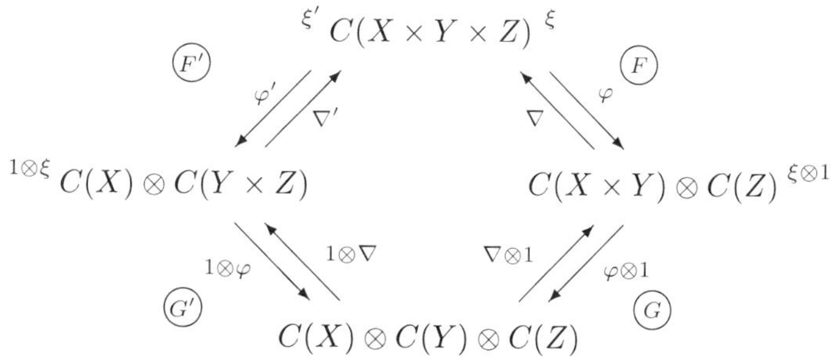

We are interested here in the special case $X = Y = Z$. We write $\varphi^3 := (1\otimes\varphi)\varphi' = (\varphi\otimes 1)\varphi$ and $\nabla^3 := \nabla'(1\otimes\nabla) = \nabla(\nabla\otimes 1)$. These well-known equations show that the two compositions of EZ-morphisms in (D.3) differ only in there chain homotopy, the other data being $C(X\times Y\times Z), \varphi^3, \nabla^3, C(X)\otimes C(Y)\otimes C(Z)$. On the right-hand side we get the chain homotopy $q = \xi + \nabla(\xi\otimes 1)\varphi$, while on the left-hand side we get $q' = \xi' + \nabla'(1\otimes\xi)\varphi'$. Hence we obtain two extensions of φ^3 to $\langle \mathrm{dgA}\rangle_1$, denoted by E resp. E'. We also have the derivation homotopy $Q : \Omega(\nabla^3)\,E \simeq 1_{\Omega C(X\times Y\times Z)}$ extending q.

We know that $K := E'\,Q$ is a derivation homotopy $E \simeq E'$ (2.74). By (2.73) we have $E = G\,F$, $E' = G'\,F'$, hence

$$K = G' \circ F' \circ Q : G\,F \simeq G'\,F' .$$

We obtain the desired natural derivation homotopy

$$\Gamma^2 := j^3 \circ K \circ \Omega C(\Delta^3) : (\psi\otimes 1)\psi \simeq (1\otimes\psi)\psi ,$$

where we need diagram (D.4) below. It is commutative except for the square $(\simeq)$ which commutes up to homotopy K. In fact the squares adjacent to $(\simeq)$ commute by naturality of F, resp. by (2.72). We also use that j is natural and associative, which is easily checked. — The last assertion of the theorem is straightforward. □

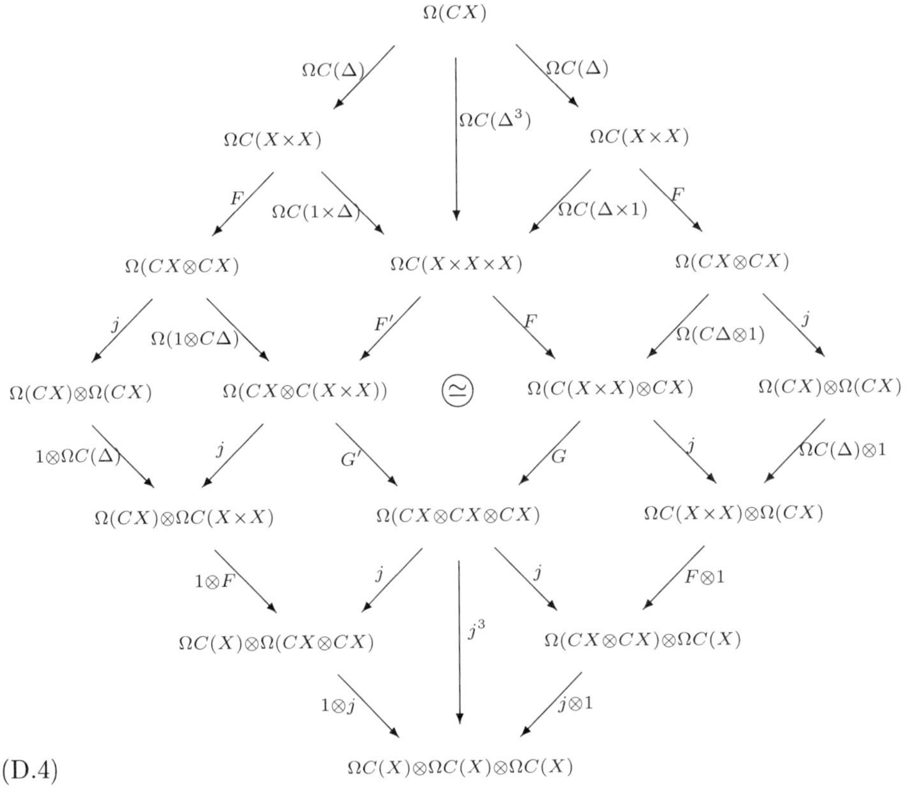

(D.4)

N° 2 A natural Hopf diagonal

We deform the natural diagonal ψ on $\Omega C(X)$ into one which is Hopf in the mild range. We use results of N° 3 and N° 4.

Let $r \geqslant 1$ be an integer. Let $p = \rho(\mathsf{K})$ be the smallest prime number not invertible in K. We put $m = rp$. In case K is a $\mathbb{Q}$-algebra we put $p = \infty$ and $m = \infty$. The *mild range* is the intervall of integers $[r, m)$. Note that $\Omega C(X)_{(m)}$ is by definition the subalgebra generated by the elements of degree $< m$, which is identified with $\Omega C(X^m)$ where X^m is the m-skeleton of the space X.

THEOREM (D.5)

For $X \in \mathcal{S}_{r+1}$ there is a natural counital diagonal χ on $\Omega C(X)$ which is Hopf on $\Omega C(X)_{(m)}$; and there is a natural homotopy $\chi \simeq \psi$. Moreover, there is a natural graded submodule $W = W_X \subset \Omega C(X)$ which freely generates $\Omega C(X)$ and such that χ is primitive on $(W \cup dW)_{<m}$.

(W natural means that for any $f : X \to Y \in \mathcal{S}_{r+1}$ the graded module W_X is carried into W_Y by $\Omega C(f)$.)

Proof. This is an immediate consequence of (D.2) using a natural version of Anick's first theorem, see (D.33). □

The rational case (D.6)

Let K be a $\mathbb{Q}$-algebra, in which case it is reasonable to put $r = 1$. Then (D.2) and Anick's first theorem (§ 4, N° 2) yields a Hopf diagonal χ homotopic to ψ on $\Omega C(X)$. This yields the result except for the naturality. The MMC–Theorem shows that $\Omega C(X), \chi$ is the universal enveloping dgH of the primitive dgL $\mathcal{M}(X) := \mathrm{Ker}\,(\overline{\chi})$. Then (by construction) W is just the image of $s_{-1}\widetilde{C}(X)$ under the canonical retraction $(\rho = p^1 \circ e^{-1})$ of $U\mathcal{M}(X) = \Omega C(X)$ onto $\mathcal{M}(X)$ given by the PBW–Theorem. It is clear that $\mathcal{M}(X) = (\mathsf{L}W, d)$.

Definition (D.7)

For $X \in \mathcal{S}_{r+1}$ let $\mathcal{M}(X)$ be the sub-dgL of $\Omega C(X)$ generated by the χ-primitives in degrees $< m$, i.e. $\mathcal{M}(X) := \mathrm{Ker}\,(\overline{\chi})_{(m)}$. This gives rise to a functor

$$\mathcal{M} : \mathcal{S}_{r+1} \to \langle \mathrm{dgL} \rangle_r$$

which we call the *cellular Lie algebra model.*

We next apply a version of the MMC–Theorem for free dg Hopf resp. Lie algebras, which holds over any ring, see N° 3.

COROLLARY (D.8)

The cellular Lie algebra model satisfies $U\mathcal{M}(X) = \Omega C(X)_{(m)}, \chi$. Moreover, $\mathcal{M}(X)$ is a free dgL generated by the natural graded submodule $W_{<m}$, i.e. $\mathcal{M}(X) = \mathsf{L}'(W_{<m}), d$. □

Remarks (D.9)

(a) It is easy to see that $\mathcal{M}(X) = [\mathcal{P}(\Omega C(X^m), \chi)]_{(m)}$. We may thus regard $\mathcal{M}$ as a functor from the category $\mathcal{S}_{r+1}^{\leqslant m}$ of $(r+1)$-reduced spaces of dimension $\leqslant m$ to the category $\mathbf{dgL}_r^{<m}$ of mild dgLs, cf. N° 4.

(b) Different natural diagonals χ as in (D.5) yield (naturally) isomorphic functors $\mathcal{M}$. This follows from (a natural version of) Anick's second theorem (D.29), cf. (D.35).

(c) Since no arbitrary choices are made in the construction of χ we have a distinguished Hopf diagonal χ in the mild range. However the construction is fairly complicated and the author has not tried to achieve an explicit formula.

N° 3 The category of dg Lie algebras (over any ring)

Recall that K is an arbitrary commutative ring. We define the category $\langle\text{dgL}\rangle$ as in § 3, N° 1, i.e. a dgL is a non-associative algebra satisfying the (graded) anti-commutativity and Jacobi identity. There is a *universal enveloping functor* U : $\langle\text{dgL}\rangle \to \langle\text{dgA}\rangle_0$ left adjoint to the forgetful functor $A \mapsto \overline{A}$. In a canonical way this becomes a functor $U : \langle\text{dgL}\rangle \to \langle\text{dgH}\rangle$ left adjoint to the primitive functor $\mathcal{P}$. There is also a functor $\mathsf{L} : \langle\text{dg}\rangle \to \langle\text{dgL}\rangle$ left adjoint to the underlying-chain-complex functor; in case V is a free chain complex $\mathsf{L}V$ is called the free dgL generated by V. Of course, $\mathsf{L}V$ is constructed as a quotient of the free magma (i.e. non-associative graded algebra) generated by V. However this construction is not used here, because only faithful dgLs are of interest.

Definition (D.10)

We call $L \in \langle\text{dgL}\rangle$ *faithful* if the canonical map $j : L \to UL$ is injective (equivalently L has a faithful representation).

There is a functor $\langle\text{dgL}\rangle \to \langle\text{faithful dgL}\rangle$ left adjoint to the embedding, namely $L \mapsto L' = \text{Im}(j)$. One checks that $U(L') = U(L)$, hence L' is indeed faithul. It is clear that the functor $V \mapsto \mathsf{L}'V = (\mathsf{L}V)' : \langle\text{dg}\rangle \to \langle\text{faithful dgL}\rangle$ is left adjoint to the underlying-chain-complex functor. Since obviously the canonical map $j : \mathsf{L}V \to \mathsf{T}V$ is a universal enveloping, the induced map $\mathsf{L}'V \to \mathsf{T}V$ is injective.

Definition (D.11)

We call a dgL *free* if it isomorphic to $(\mathsf{L}'V, d)$ for some free graded module V and any differential d.

Notice that $\mathsf{L}'V$ is just the sub-dgL of $\mathsf{T}V$ generated by V. This preferable definition of $\mathsf{L}'V$ plainly shows that $\mathsf{L}'V = \bigoplus_{k=1}^{\infty} \mathsf{L}'^k V$ where $\mathsf{L}'^k V = [\mathsf{L}'^{k-1} V, V]$ and $\mathsf{L}'^1 V = V$. Notice that any graded map $\partial : V \to \mathsf{L}'V$ extends uniquely to a derivation ∂ on $\mathsf{L}'V$. One checks that when (V, ∂) is a chain complex then $\mathsf{L}'(V, \partial) = (\mathsf{L}'V, d)$ where on the right-hand side we regard V as a graded module and d is the unique extension of ∂. These facts follow easily from the corresponding properties of the tensor algebra. (There are analogous properties of $\mathsf{L}V$ but here one has to consult its rather difficult construction.)

Definition (D.12)

For $L \in \langle\text{dgL}\rangle$ let the *Lie filtration* of UL be the increasing filtration where F_n is the sub module generated by all products of k or less elements in $j(L)$, i.e. $F_0 = \mathsf{K}$ and $F_n = F_{n-1} \cdot (\mathsf{K} \oplus j(L))$.

Then $\bigoplus_n \text{gr}_n(UL)$ is an augmented dgA in a canonical way, and an easy computation shows that this is a commutative dgA. Recall that the symmetric algebra $\mathsf{S}L \in \langle\text{dgH}\rangle$ is the universal algebra of L considered as an abelian dgL. Then $\mathsf{S}L = \bigoplus_{k=0}^{\infty} \mathsf{S}^n L$ where this decomposition comes from that of the tensor algebra of which it is a quotient. It is clear that $F_n(\mathsf{S}L) = \bigoplus_{k=0}^n \mathsf{S}^k L$. The obvious chain map $L \to \mathrm{gr}_1(UL)$ induces the *canonical* (surjective) cdgA-map $\mathsf{S}(L) \to \mathrm{gr}(UL)$.

THEOREM (D.13)

Let $\frac{1}{2} \in \mathsf{K}$. Let L be a faithful dgL which is degreewise free. Then there is a filtered graded module isomorphism $\gamma : \mathsf{S}L \to UL$ compatible with the diagonals.

The associated graded of γ is the canonical algebra map. □

Remark (D.14)

Notice that γ is *not* natural and *not* differential. As pointed out by [CMN], the proof of this PBW–Theorem (given in [Jo]) goes through if the assumption on L being faithful is replaced by the (plainly weaker) condition

$(*)$ $\quad [[x,x],x] = 0$ for $x \in L^{\mathrm{odd}}$.

(Note that the Jacobi identity implies that $3[[x,x],x] = 0$ for $x \in L^{\mathrm{odd}}$). Hence if L is K-free and satisfies $(*)$ then L is faithful provided $\frac{1}{2} \in \mathsf{K}$. This is generalized immediately to the case where $\mathsf{K} \to \mathsf{K}[\frac{1}{2}]$ is injective which means K is 2-torsionfree. Note that $\mathsf{K}' \otimes UL = U_{\mathsf{K}'}(\mathsf{K}' \otimes L)$ for any ring extension $\mathsf{K}' \supset \mathsf{K}$.

THEOREM (D.15)

Let $\frac{1}{2} \in \mathsf{K}$. When V is a free graded module so is $\mathsf{L}'V$.

Proof. Let $R = \mathbb{Z}[\frac{1}{2}]$ and note that by assumption $\mathsf{K} = \mathsf{K} \otimes R$ where the tensor product is over R (or over $\mathbb{Z}$ which amounts to the same). Thus $V \cong \mathsf{K} \otimes W$ where W is R-free. Since $j : L'_R W \hookrightarrow \mathsf{T}_R W$ is injective and R is hereditary, $L'_R W$ is R-free. Hence j has an R-linear retraction (D.13), so that in the obvious sequence $\mathsf{K}\otimes \mathsf{L}_R W \twoheadrightarrow \mathsf{K}\otimes \mathsf{L}'_R W \hookrightarrow \mathsf{K}\otimes \mathsf{T}_R W$ the last map is injective. Now $\mathsf{K}\otimes \mathsf{L}_R W = \mathsf{L}_{\mathsf{K}} V$ and $\mathsf{K}\otimes \mathsf{T}_R W = \mathsf{T}_{\mathsf{K}} V$ by adjoint functor arguments, hence $\mathsf{K}\otimes \mathsf{L}'_R W = \mathsf{L}'_{\mathsf{K}} V$ yielding the assertion. □

Remark (D.16)

It is possible to describe $\mathsf{L}V$ explicitly in terms of "basic products" of basis elements of V, see [Hi]. This holds over any ring, but it suffices to consider the integral case $\mathsf{K} = \mathbb{Z}$. Hilton shows in particular that $\mathsf{L}V$ is (in each degree) a direct sum of a free module X^1, a mod 2 free module X^2 and a mod 3 free module X^3. The basis elements of X^2 resp. X^3 are of the form $[y,y]$ where $y \in (\mathsf{L}V)^{\mathrm{even}}$ resp. $[[x,x],x]$ where $x \in (\mathsf{L}V)^{\mathrm{odd}}$. — Of course these elements vanish in $\mathsf{L}'V$, and in fact one may show that $\mathsf{L}V/(X^2 \oplus X^3) = \mathsf{L}'V$ provided $\frac{1}{2} \in \mathsf{K}$ (or at least $\mathsf{K} \subset \mathsf{K}[\frac{1}{2}]$). As in the proof above this reduces to the case $\mathsf{K} = R$ which follows from the observation that the quotient graded Lie algebra $\mathsf{L}V/(X^2 \oplus X^3)$ is torsionfree and hence satisfies (D.14) $(*)$. By similar arguments one has $\mathsf{L}V = \mathsf{L}'V$ provided $\frac{1}{2}, \frac{1}{3} \in \mathsf{K}$.

Of great importance is the following version of the PBW–Theorem which is a generalization of (4.53). Recall that $p = \rho(\mathsf{K})$ denotes the smallest prime not invertible in K. Notice that the map $e : F_{p-1}(\mathsf{S}L) \to F_{p-1}(UL)$ given by the exponential formula is well-defined for any L. To say that it is compatible with diagonal means that e is a coalgebra map into UL. (In case L is as in (D.13) that theorem shows $F_k UL$ is a direct summand of UL and the reduced diagonal restricts to $\overline{\Delta} : F_k \to F_{k-1} \otimes F_{k-1}$, thus $F_k UL$ is a sub-dgC of UL.)

PBW–THEOREM (D.17)

Let L be a faithful dgL which is degreewise free. The map $e : F_{p-1}(\mathsf{S}L) \to F_{p-1}(UL)$ is a filtered dg isomorphism compatible with the diagonals.

Let L be an r-reduced faithful dgL. Then $e : \mathsf{S}L \to UL$ is a well-defined dg isomorphism in degrees $< rp$ compatible with the diagonals.

Proof. One checks by computation that indeed e is compatible with filtration, differential and diagonal. To show e bijective is trivial in case $p = 2$. In case $p > 2$ (i.e. $\frac{1}{2} \in \mathsf{K}$) the first assertion follows from (D.13) since $\mathrm{gr}_{<p}(e)$ is the obvious map mentioned above. The second follows immediately when L is faithful *and* degreewise free. Now if L is just faithful there is a surjection $\mathsf{L}'V \twoheadrightarrow L$ with V a free graded module. Then as in (3.54) one reduces to the case where $L = \mathsf{L}'(V)$. In this case L is degreewise free by (D.15). □

Remark (D.18)

The second part is of main interest here. Using the previous remark one proves similarly that this assertion is actually true for *any* dgL.

Anick's work. The remainder of this appendix is a report of Anick's work [An], with some improvements. The story begins with an inspired investigation of the symmetric algebra $\mathsf{S}L$ where L is a chain complex. Recall that $A = \mathsf{S}L \in \langle \mathrm{dgH} \rangle$ where the diagonal Δ is unique with the property $L \subset \mathcal{P}A$. An easy computation shows:

$$\text{(D.19)} \qquad \overline{\Delta}(x_1 \dots x_k) = \sum_{i=1}^{k-1} \sum_{\sigma} \pm x_{\sigma(1)} \cdots x_{\sigma(i)} \otimes x_{\sigma(i+1)} \cdots x_{\sigma(k)}$$

for any $k \geqslant 1$ and $x_1, \dots, x_k \in L$. Here σ runs through the set of all $(i, k-i)$-shuffles, i.e. permutations of $\{1, \dots, k\}$ that preserve order on $\{1, \dots, i\}$ and $\{i+1, \dots, k\}$.

Let $\bigoplus_{k \geqslant 1} \overline{A}^{[k]}$ be the canonical direct sum decomposition, i.e. $\overline{A}^{[k]} = \mathsf{S}^k L$. Note that $F_n \overline{A} = \bigoplus_{k \leqslant n} \overline{A}^{[k]}$ is induced by the Lie filtration of A. There is an induced filtration resp. decomposition of $\overline{A}^{\otimes 2}$ given by

$$\begin{aligned} F_n(\overline{A} \otimes \overline{A}) &= \textstyle\sum_{i=1}^{n-1} F_i \overline{A} \otimes F_{n-i} \overline{A} \\ &= \textstyle\bigoplus_{k \leqslant n} \bigoplus_{i=1}^{k-1} \overline{A}^{[i]} \otimes \overline{A}^{[k-i]} \end{aligned}$$

$\subset F_{n-1}\overline{A} \otimes F_{n-1}\overline{A}$. Analogously for $\overline{A}^{\otimes 3}$.

Let $\overline{\overline{\Delta}} = (1 - \tau, \overline{\Delta} \otimes 1 - 1 \otimes \overline{\Delta})$ regarded as a chain map $\overline{A}^{\otimes 2} \to \overline{A}^{\otimes 2} \oplus \overline{A}^{\otimes 3}$. Observe that $\overline{\Delta}$ and $\overline{\overline{\Delta}}$ preserve filtration (even decomposition). Regard the following diagram.

$$(*) \qquad 0 \to L \xrightarrow{\ j\ } F_{p-1}\overline{A} \underset{\mu}{\overset{\overline{\Delta}}{\rightleftarrows}} F_{p-1}(\overline{A}^{\otimes 2}) \underset{\lambda}{\overset{\overline{\overline{\Delta}}}{\rightleftarrows}} F_{p-1}(\overline{A}^{\otimes 2} \oplus \overline{A}^{\otimes 3})$$

LEMMA (D.20)

The sequence of solid arrows is exact. There are (canonical) chain maps μ, λ *which are splittings in the sense that* $\overline{\Delta}\mu + \lambda\overline{\overline{\Delta}} = 1$. *Moreover* $\mathrm{Im}(\mu) \subset \overline{A}^2$.

Hence: $\overline{\Delta}\,\mu\,\overline{\Delta} = \overline{\Delta}$, $\overline{\overline{\Delta}}\,\lambda\,\overline{\overline{\Delta}} = \overline{\overline{\Delta}}$ *and* $1 - \mu\,\overline{\Delta} : F_{p-1}\overline{A} \to L$ *is the canonical retraction.*

Proof. Observe that $\overline{\overline{\Delta}}\,\overline{\Delta} = 0$ is equivalent to the fact that Δ is co-commutative and -associative. Exactness is a consequence of the splitting property. We define μ and λ_1, λ_2 (the two components of λ) by giving their restrictions to $\overline{A}^{[i]} \otimes \overline{A}^{[k-i]}$ resp. $\overline{A}^{[i]} \otimes \overline{A}^{[j]} \otimes \overline{A}^{[k-i-j]}$ $(0 < i, j, i+j < k;\ k \leqslant p-1)$, denoted by $\mu^{k,i}$ and $\lambda_1^{k,i}$, resp. $\lambda_2^{k,i,j}$. Let

$$\begin{aligned} \mu^{k,1} &= \tfrac{1}{k}\mathrm{mult}_A\,, \\ \mu^{k,i} &= 0 \text{ for } i \geqslant 2\,, \\ \lambda_1^{k,i} &= -\tfrac{i}{k}\tau\,, \\ \lambda_2^{k,i,j} &= \tfrac{i+j}{k}(1+\tau)(\mu \otimes 1)\,. \end{aligned}$$

The verification of the desired properties presents no difficulty. □

Remark (D.21)

Let $\frac{1}{2} \in \mathsf{K}$ and L be faithful and free in each degree. By the PBW–Theorem we have the canonical filtered differential isomorphism $e : F_{p-1}\overline{\mathsf{S}L} \xrightarrow{\approx} F_{p-1}\overline{UL}$. Since the filtration of UL is splittable (i.e. consists of direct summands) by (D.13), we have $\mathrm{gr}(e \otimes e) = \mathrm{gr}(e) \otimes \mathrm{gr}(e)$ and it follows that $e^{\otimes 2} : F_{p-1}(\overline{\mathsf{S}L}^{\otimes 2}) \to F_{p-1}(\overline{UL}^{\otimes 2})$ is a filtered differential isomorphism, and analogously for $e^{\otimes 3}$. Since e is compatible with diagonals it also follows that the maps $\overline{\Delta} : \overline{UL} \to \overline{UL}^{\otimes 2}$ and $\overline{\overline{\Delta}} : \overline{UL}^{\otimes 2} \to \overline{UL}^{\otimes 2} \oplus \overline{UL}^{\otimes 2}$ restrict to filtration $p-1$.

Hence Lemma (D.20) holds with $A = UL$. Here the maps μ, λ are unique with the property that e, $e^{\otimes 2}$ and $e^{\otimes 2} \oplus e^{\otimes 3}$ constitute an isomorphism between [diagram $(*)$ with $A = \mathsf{S}L$] and [diagram $(*)$ with $A = UL$]. Note that the *canonical* retraction of $F_{p-1}(UL)$ onto L is the composite of e^{-1} followed by the canonical retraction $p^1 : \mathsf{S}L \to L$.

We leave to the reader the exact formulation of these results. Below we give another version which is sufficient for our purposes.

Der(L)-linearity (D.22)

The derivations of a dgL L form a sub-graded Lie algebra $\mathrm{Der}(L)$ of the graded algebra $\mathrm{End}(L), \circ$ of graded maps $L \to L$. Regard L as a (left) $\mathrm{Der}(L)$-module in the obvious way. There is an obvious definition of tensor product (resp. direct sum) of (left) $\mathrm{Der}(L)$-modules. Thus $\mathsf{T}L$ is a $\mathrm{Der}(L)$-module. One checks that $\mathsf{S}L$ is a quotient module, and that its canonical increasing filtration resp. decomposition is one of sub $\mathrm{Der}(L)$-modules. Also UL is a quotient $\mathrm{Der}(L)$-module of $\mathsf{T}L$, and the Lie filtration is one of $\mathrm{Der}(L)$-modules. Observe that any $\delta \in \mathrm{Der}(L)$ acts as a derivation and coderivation on UL resp. $\mathsf{S}L$, $\mathsf{T}L$ (i.e. mult and Δ are $\mathrm{Der}(L)$-linear). Only two types of L-derivations are of interest here:

(a) The differential d_L is an L-derivation. Its action on UL is given by $d_L x = d(x)$, the differential of UL. Analogously for $\mathsf{S}L$, $\mathsf{T}L$.

(b) Any $b \in L$ yields the L-derivation $\operatorname{ad}(b) = [b, -]$. Thus any $\operatorname{Der}(L)$-module is in particular an L-module. The action of $\operatorname{ad}(b)$ on $UL =: A$ is given by $\operatorname{ad}(b)\, x = [b, x]$, the commutator bracket in A. Hence $A^{\otimes 2}$ is an L-module via $\operatorname{ad}(b)\, y = [b \otimes 1 + 1 \otimes b, y]$; and similarly for $A^{\otimes 3}$.

It is readily verified that all maps in diagram $(*)$ with $A = \mathsf{S}L$ (in particular μ and λ) are $\operatorname{Der}(L)$-linear. Fix $r \geqslant 1$, put $m = rp$ and consider the following diagram. There is an obvious notion of $\operatorname{Der}(L)$-linear map defined in degrees $< k$, cf. (D.24) (a).

$$(**) \qquad 0 \to L \xrightarrow{\ j\ } \overline{UL} \underset{\mu}{\overset{\overline{\Delta}}{\rightleftarrows}} \overline{UL}^{\otimes 2} \underset{\lambda}{\overset{\overline{\overline{\Delta}}}{\rightleftarrows}} \overline{UL}^{\otimes 2} \oplus \overline{UL}^{\otimes 3}$$

THEOREM (D.23)

Let L be r-reduced. In degrees $< m$, the sequence of solid $\operatorname{Der}(L)$*-maps is exact and there are canonical* $\operatorname{Der}(L)$*-linear splittings μ, λ.*

One can extend μ, λ to natural $\operatorname{Der}(L)$*-linear maps in degrees $\leqslant m$.*

Moreover $\operatorname{Im}(\mu) \subset \overline{UL}^{2}$. □

Proof. One checks that $e : F_{p-1}\mathsf{S}L \to F_{p-1}UL$ is $\operatorname{Der}(L)$-linear. By the second part of the PBW–Theorem (D.17), the restriction $e : \overline{\mathsf{S}L} \to \overline{UL}$ is an isomorphism in degrees $< m$ and trivial in degree 0, thus $e^{\otimes 2}$ and $e^{\otimes 3}$ are isomorphisms in degrees $\leqslant m$. — We define μ, λ in such a way that they commute via e, $e^{\otimes 2}$ and $e^{\otimes 2} \oplus e^{\otimes 3}$ with the corresponding maps of (D.20). Then we are reduced to the abelian case, where it remains to show that in the corresponding sequence for $\mathsf{S}L = A$ there are $\operatorname{Der}(L)$-linear extensions of μ resp. λ in degrees $\leqslant m$ whose image is contained in F_{p-1}. This is indeed the case, put $\mu = 0$ on $(\overline{A}^{[i]} \otimes \overline{A}^{[p-i]})_m$, $i = 1, \dots, p-1$, and analogously for λ. □

Remarks (D.24)

(a) Let M, N be $\operatorname{Der}(L)$-modules. Let a mapping $f : M \to N$ (of degree 0) be defined in degrees $\leqslant n$. Then f is said to be $\operatorname{Der}(L)$-linear if $\delta\, f(x) = f(\delta\, x)$ for all $\delta \in \operatorname{Der}(L)$, $x \in M : |x|, |\delta| + |x| \leqslant n$.

(b) We also see that the canonical retraction $\rho : UL \to L$ in degrees $< m$ is $\operatorname{Der}(L)$-linear ($\rho = 1 - \mu\overline{\Delta} = p^1 \circ e^{-1}$). In case $p > 2$ we can extend ρ to a retraction chain map in degrees $\leqslant m$. This is needed for the general version of the Aubry-Lemaire theorem (D.30), which is stated but not proved in [An]. In fact, using the PBW–Theorem (D.13) for the last time, we have $(UL)_m \approx F_{p-1}U(L)_m \oplus \mathsf{S}^p(L_r)$. Since ρ is already defined as a chain map on the first summand and $\mathsf{S}^p(L_r)$ is a module of cycles, putting $\rho = 0$ on the second yields the desired chain map. Notice that we need here that L is faithful and degreewise free.

Recall that r is fixed, $p = \rho(\mathsf{K})$ and $m = rp$. The following theorem is a version of the MMC–Theorem. In fact it implies an equivalence of categories between mild dgHs and mild dgLs. Note that if L is mild (i.e. freely generated in degrees $< m$) so is UL and $L = (\mathcal{P}UL)_{(m)}$ by (D.23). For variants see remark below.

Theorem (D.25)

Let $A \in \langle \mathrm{dgH} \rangle_r$ be free with $A = A_{(m)}$. We have the canonical isomorphism $UL = A$ where $L = (\mathcal{P}A)_{(m)}$. This is a free dgL. □

Hence if $A = (\mathsf{T}V, d)$ then $L = (\mathsf{L}'W, d)$. Notice that necessarily $W \approx V$. Given V the proof below shows that we can choose $W = \rho(V)$ where $\rho : UL \to L$ is the canonical retraction.

Proof. Let $A = (\mathsf{T}V, d, \chi)$. Notice that $A_{(q)} = (\mathsf{T}V_{<q}, d, \chi)$ is a sub-dgH. In case $A = A_{(q)}$ with $q \leqslant 2r$ we have $d(V) \subset V$ and $\overline{\chi}(V) = 0$ by reason of degree. Hence $L = (\mathsf{L}'V, d)$ satisfies $UL = A$ and it is clear by exactness of $(**)$ in degrees $< m$ that $L = (\mathcal{P}A)_{(m)}$. This completes the proof in case $p = 2$. — Now fix $q < m$ and by induction suppose the assertion of the theorem for free dgHs generated in degrees $< q$. Let A be generated in degrees $\leqslant q$. Apply (D.23) to $\check{L} = (\mathcal{P}\, A_{(q)})_{(m)}$ and note that by the inductive hypothesis $U\check{L} = A_{(q)}$. Put $L = (\mathcal{P}A)_{(m)}$ which coincides with $\check{L}$ in degrees $< q$. Since $i : \overline{A}_{(q)} \hookrightarrow \overline{A}$ is the identity mapping in degrees $< q$ and A is reduced, it follows that $i^{\otimes 2}$ and $i^{\otimes 2} \oplus i^{\otimes 3}$ are identity mappings in degrees $\leqslant q$. Hence we have the following diagram

$$\text{(D.26)} \qquad L \hookrightarrow \overline{A} \underset{\mu}{\overset{\overline{\chi}}{\rightleftarrows}} \overline{A}^{\otimes 2} \underset{\lambda}{\overset{\overline{\overline{\chi}}}{\rightleftarrows}} \overline{A}^{\otimes 2} \oplus \overline{A}^{\otimes 3}$$

where consecutive maps from left to right compose to 0 and are exact in degrees $\leqslant q$, where μ, λ are defined and satisfy the splitting relation $\overline{\chi}\,\mu + \lambda\,\overline{\overline{\chi}} = 1$. This implies that $\rho = 1 - \mu\overline{\chi} : \overline{A} \to L$ is a retraction onto L in degrees $\leqslant q$. By construction $\mathrm{Im}(\mu) \subset \overline{A}_{(q)}$ hence in degree q this image consists of decomposable elements. It follows that ρ is injective on V_q and we get the free module $W_q = \rho(V_q) \subset L$. It also follows that the inclusion induces a dgH-isomorphism $\iota : \mathsf{T}W \xrightarrow{\approx} A$ where $W_{<q} \subset L$ is such that $\check{L} = \mathsf{L}'(W_{<q})$ (inductive hypothesis) and $W_{>q} = 0$. The restriction of ι is an injective map $\mathsf{L}'W \to L$. Since the diagram above and $(**)$ in case of $\mathsf{L}'W$ are exact in degrees $\leqslant q$, we see that $\mathsf{L}'W \xrightarrow{\approx} L$ in these and thus in all degrees. From $U(\mathsf{L}'W) = \mathsf{T}W$ it now follows that $UL = A$ and the inductive step is complete. In case $m < \infty$ this induction proves the theorem in a finite number of steps; in case $m = \infty$ it follows immediately. □

Remark (D.27)

For any $A \in \langle \mathrm{dgH} \rangle_r$ one still has $UL = A$ in degrees $< m$. In fact, the argument of the inductive step shows the surjectivity of the desired map, and its bijectivity follows by a standard trick (3.64) $(*)$ using that $\mathcal{P}(UL) = L$ in degrees $< m$ by (D.23). — In case K is an integral domain of characteristic 0 the latter holds in all degrees (see [CMN | §2]), hence $UL = A$ when $A = A_{(m)}$.

N° 4 Anick's theorems and naturality

Fix $r \geqslant 1$. Recall $p = \rho(\mathsf{K})$ is the smallest prime number not invertible in K, and $m = rp$. The mild range is the intervall of integers $[r, m)$. We present the theorems of Anick and Aubry - Lemaire as given in [An]. We formulate a natural version of Anick's first theorem which is used in N° 2. We sketch the proof exhibiting all details necessary to show that Anick's construction can be made natural. We also intend to make the ideas accessible and give some arguments that are missing in [An].

An object $A \in \langle \text{dgA} \rangle_r$ resp. $\langle \text{dgHh} \rangle_r$ resp. $\langle \text{dgL} \rangle_r$ is called *mild*, if A is free and $A = A_{(m)}$ i.e. it is generated in the mild range. In the last two cases we obtain categories $\mathbf{dgHh}_r^{<m} = \langle \text{mild dgHh} \rangle$ resp. $\mathbf{dgL}_r^{<m} = \langle \text{mild dgL} \rangle$. Recall that any free (in particular any mild) dgL is faithful. Recall that a map in $\mathbf{dgHh}_r^{<m}$ is a dgA-map commuting up to derivation homotopy with any representants of the homotopy diagonals.

ANICK'S FIRST THEOREM (D.28)
Let $(A, \Psi) \in \mathbf{dgHh}_r^{<m}$. Then there is a Hopf diagonal χ on A with $[\chi] = \Psi$. □

Note that $(A, \chi) = UL$ with $L \in \mathbf{dgL}_r^{<m}$ by (D.25).

ANICK'S SECOND THEOREM (D.29)
Let L, $M \in \langle \text{dgL} \rangle_r$ where L is mild. Let $f : UL \to UM$ be a dgHh-map. Then $f \simeq U(\ell)$ for some dgL-map $\ell : L \to M$ which is unique up to homotopy.

Moreover, if f is an isomorphism so is ℓ. □

The uniqueness of ℓ up to homotopy is actually a consequence of the theorem below, which is a straightforward generalization of (4.57). Here we define (following Anick) the natural equivalence relation ($\simeq$) between maps of $\langle \text{dgL} \rangle_r$ whose common source is mild, by means of the Tanré cylinder object. (The definition is the same as in the rational case).

AUBRY - LEMAIRE THEOREM (D.30)
Let $f, g : L \rightrightarrows M \in \langle \text{dgL} \rangle_r$ and let L be mild. Then $f \simeq g$ iff $U(f) \simeq U(g)$. □

Combining these theorems yields ...

ANICK'S EQUIVALENCE (D.31)
The functor $U : \mathbf{dgL}_r^{<m} \to \mathbf{dgHh}_r^{<m}$ induces an equivalence of categories

$$U : \mathbf{dgL}_r^{<m}/_{\simeq} \xrightarrow{\approx} \mathbf{dgHh}_r^{<m}/_{\simeq} .$$

□

We only discuss here the proof of Anick's first theorem, the proof of the second is similar. To formulate the naturality of Anick's construction, we need the technical definition below. First some preliminary remarks.

We are going to prove a slightly stronger version of Anick's first theorem: Given $(A, \Psi) \in \langle \text{dgHh} \rangle_r$ where A is free (but not necessarily mild) there is a diagonal χ with $[\chi] = \Psi$ and such that χ is Hopf in the mild range, i.e. $A_{(m)}$ with the restriction of χ is a dgH.

Note that we can always represent Ψ by a *counital* diagonal. This follows from a standard lifting argument using that $(\pi_1, \pi_2) : A \otimes A \to A \times A$ is surjective, hence a fibration in the model category $\langle \text{dgA} \rangle_0$.

Definitions (D.32)

Let A be a free reduced dgA and $V \subset A$ a freely generating graded submodule; let ψ be a counital diagonal on A with derivation homotopies $G_1 : \psi \simeq T\psi$, $G_2 : (\psi \otimes 1)\psi \simeq (1 \otimes \psi)\psi$.—We call A together with such a quadruplet (V, ψ, G_1, G_2) an *explicit dgHh*.

Let $f : A \to \tilde{A}$ be a dgA-map where both A, $\tilde{A}$ are endowed with explicit dgHh-structure. Then f *preserves* this structure if $f(V) \subset \tilde{V}$, $(f \otimes f)\psi = \tilde{\psi} f$, $(f \otimes f)G_1 = \tilde{G}_1 f$, and $(f \otimes f \otimes f)G_2 = \tilde{G}_2 f$.

THEOREM (D.33)

Let $r \geqslant 1$. Let A be an r-reduced free dgA, endowed with explicit dgHh-structure (V, ψ, G_1, G_2). There exists a "new" explicit dgHh-structure (W, χ, J_1, J_2) and a homotopy $F : \psi \simeq \chi$, such that χ is Hopf on $A_{(m)}$ and primitive on $(W \cup dW)_{<m}$.

This construction is natural in the following sense: if $f : A \to \tilde{A}$ preserves the given explicit dgHh-structure, then f also preserves the new structure; moreover $\tilde{F} f = (f \otimes f)\, F$.

Remark. In case $\mathsf{K} \supset \mathbb{Q}$ we have $m = \infty$, thus (A, χ) is a dgH, χ is primitive on W (hence on dW), and we can choose $J_1 = 0,\ J_2 = 0$.

Sketch of proof. It is clear that ψ is primitive on $A_{<2r}$. Thus in case $p = 2$ the assertion is trivial. Let $p > 2$. By induction suppose ψ is already Hopf on $A_{(q)}$ and primitive on $(V \cup dV)_{<q}$ where $q < m$. For the inductive step we have to prove the assertion of the theorem with m replaced by $q + 1$. (□)

Inductive step. By (D.25) we have $A_{(q)} = U(L)$ as dgHs where $L = \operatorname{Ker}(\overline{\psi})_{(q)}$. As in the proof of that theorem we obtain a natural diagram

(D.34)
$$\overline{A} \underset{\mu}{\overset{\overline{\psi}}{\rightleftarrows}} \overline{A}^{\otimes 2} \underset{\lambda}{\overset{\overline{\overline{\psi}}}{\rightleftarrows}} \overline{A}^{\otimes 2} \oplus \overline{A}^{\otimes 3}$$

where μ, λ are $\operatorname{Der}(L)$-maps defined in degrees $\leqslant q$ where they satisfy $\overline{\psi}\,\mu + \lambda\,\overline{\overline{\psi}} = 1$ and $\operatorname{Im}(\mu) \subset \overline{A}^2$. (Notice that $\overline{\overline{\psi}}\,\overline{\psi} = 0$ in degrees $< q$.) Here we need a bit more: going back to (D.23) we see that it is possible to extend μ, λ in degree $q + 1$, so that these extensions are still natural chain maps and $\operatorname{Im}(\mu) \subset \overline{A}^2$. [There is one difficulty here when $r = 1$, since in this case i^2 is not the identity mapping in degree $q + 1$ where $\overline{A}^{\otimes 2} = \overline{A}_{(q)}^{\otimes 2} \oplus (A_1 \otimes V_q) \oplus (V_q \otimes A_1)$. Hence we may extend μ and λ by putting $\mu = \frac{1}{4}\operatorname{mult}_A(1 + \tau)$ and $\lambda_1 = -\frac{1}{2}\tau$ on $(A_1 \otimes V_q) \oplus (V_q \otimes A_1)$. It is easy to verify using $d(V_q) \subset L$ that these extensions are chain maps.]

Now define a diagonal χ on A and a derivation homotopy $H : \psi \simeq \chi$ as follows. On $V_{\leqslant q}$ put $\chi = \psi - \lambda\,\overline{\overline{\psi}}\,\overline{\psi}$ and extend this uniquely to a dgA-map on $A_{(q+1)}$. On $V_{\leqslant q}$ put $H = \lambda\,\overline{\overline{\psi}}\,\lambda\,\pi\,(G_1, G_2)$ and extend this uniquely to a (ψ, χ)-derivation on $A_{(q+1)}$. [Here $\pi : A^{\otimes 2} \oplus A^{\otimes 3} \to \overline{A}^{\otimes 2} \oplus \overline{A}^{\otimes 3}$ denotes the chain map $(1 - \eta\,\varepsilon)^{\otimes 2} \oplus (1 - \eta\,\varepsilon)^{\otimes 3}$.]

Using that μ, λ are L-linear, $\psi = \chi$ in degrees $< q$ and $d(V_q) \subset L$ one computes that $DH = \psi - \chi$ on $V_{\leqslant q}$ hence on $A_{(q+1)}$. We apply the HEP (see below) of $A_{(q+1)} \hookrightarrow A$ to obtain extensions H, χ defined on A. Putting $H = 0$ on $V_{>q}$ the constructions of H, χ are natural.

It is not difficult to see that χ is a counital diagonal which is Hopf on $A_{(q+1)}$ (observe that $\overline{\chi} = \overline{\psi}\,\mu\,\overline{\psi}$ on $V_{\leqslant q}$). Using (D.36) to add or substract homotopies $A \to A \otimes A$ we obtain certain homotopies J_1, J_2 for χ, as required for an explicit dgHh-structure.

We construct a freely generating graded submodule W, such that χ is primitive on $W_{\leqslant q}$ and $d(W_{q+1})$. The first part of the (next) inductive step yields a certain chain map $\mu : \overline{A}^{\otimes 2} \to \overline{A}$ in degrees $\leqslant q{+}1$ (it agrees with the μ above in degrees $\leqslant q$) satisfying $\overline{\chi} = \overline{\chi}\,\mu\,\overline{\chi}$ on $\overline{A}_{\leqslant q}$. Put $W_{\leqslant q+1} = (1 - \mu\,\overline{\chi})(V_{\leqslant q+1})$ and $W_{>q+1} = V_{>q+1}$. The desired properties are easily verified, further note that $W_{<q} = V_{<q}$.

The construction of the "new" explicit dgHh-structure (W, χ, J_1, J_2) is clearly natural in the required sense. This completes the inductive step. $(\Box)$

In case $\rho(\mathsf{K}) < \infty$ we thus obtain the desired explicit dgHh-structure after a finite number of steps. It remains to note that the homotopy $F : \psi \simeq \chi$ is obtained by adding the corresponding homotopies constructed in each inductive step.

In case $\rho(\mathsf{K}) = \infty$, however, some more work has to be done. Since infinitely many homotopies cannot be added in general, we would like to know (returning to the inductive step) that $H = 0$ on $V_{<q-1}$.—This is easily verified if we assume, as additional inductive hypothesis, that $\pi(G_1, G_2)(V_{<q-1}) \subset \mathrm{Im}(1 - \overline{\overline{\psi}}\lambda)$. We have to verify the corresponding condition for J_1, J_2 which we construct now more carefully. Since ψ is primitive and $H = 0$ on $V_{<q-1}$, we may apply (D.36) with $k = q - 1$ and then these homotopies satisfy

$$\begin{aligned} J_1 &= -H + J_1 + \tau H \\ J_2 &= -(\chi \otimes 1)H - (H \otimes 1)\psi + J_2 + (1 \otimes H)\psi + (1 \otimes \chi)H \end{aligned}$$

on $V_{<q}$. This easily implies $\pi(J_1, J_2)(V_{<q}) \subset \mathrm{Im}(1 - \overline{\overline{\chi}}\lambda)$ completing the (modified) inductive step. $(\Box)$

Summarizing we have a sequence of counital diagonals $\psi = \chi^0 \simeq \chi^1 \simeq \chi^2 \simeq \dots$ where χ^k is Hopf on $A_{(k+1)}$; moreover $\chi^n = \chi^{n+1} = \dots$ on $A_{\leqslant n}$, so that we have a "limit" diagonal χ. We also know that $H^k : \chi^{k-1} \simeq \chi^k$ is zero on $A_{(k-1)}$; adding homotopies (naturally) as in (D.36), we obtain $F^k : \psi \simeq \chi^k$, such that $F^{n+2} = F^{n+3} = \dots$ on $A_{\leqslant n}$. Hence we get $F : \psi \simeq \chi$ as desired. This completes the proof of the theorem. $\Box$

Remark (D.35)

Also the constructions implicitly given by theorems (D.29)–(D.30) are natural in the obvious sense. This is seen by inspection of the proofs given in [An], [AL].

In the proof above two natural versions of standard homotopy theoretic constructions in the model category $\langle \mathrm{dgA} \rangle_0$ are used. Here A is a free dgA with a distinguished freely generating graded submodule $V \subset A$.

The first construction is related with the *homotopy extension property* (*HEP*) of the inclusion $A_{(k)} \hookrightarrow A$ where $k \geqslant 0$ (note that $A_{(0)} = \mathsf{K}$). This means ...

(HEP) given $\psi : A \to B$, $\chi : A_{(k)} \to B \in \langle \mathrm{dgA} \rangle_0$ and $H' : \psi_{(k)} \simeq \chi'$, there is $H : \psi \simeq \chi$ with $H_{(k)} = H'$ and $\chi_{(k)} = \chi'$.

We claim that there is a natural way to construct H and χ. On V_k, define $H = 0$ and $\chi = \psi - d\,H - H\,d$ $(= \psi - H\,d)$. Extend χ to a dgA-map, and H to a (ψ, χ)-derivation on $A_{(k+1)}$. Then $D\,H = \psi - \chi$ on $A_{(k+1)}$ and thus $H : \psi_{(k+1)} \simeq \chi$. Proceeding inductively, we obtain an extension of χ to A, and a derivation homotopy $H : \psi \simeq \chi$. We have $H|_V = 0$ by which H and χ are uniquely determined.

The second construction is the addition (or substraction) of derivation homotopies, one of which being zero on $A_{(k)}$. The assertion is stated more precisely in the next lemma. Except for the naturality, it is stated (without proof) in [An|5.1], but there the hypothesis that F (or G) is zero on $A_{(k)}$ is missing. In the absolute case (i.e. $k = 0$) a natural addition of homotopies, by means of explicit formulas, is given in [B | II.17].

LEMMA (D.36)

Let $f, g, h : A = (\mathsf{T}V, d) \to B \in \langle \mathrm{dgA} \rangle_0$ *and* $F : f \simeq g$, $G : g \simeq h$. *Suppose* $f = g = h$ *on* $A_{(k+1)}$ *and* $F = 0$ *(or* $G = 0$*) on* $A_{(k)}$.

Then there is a derivation homotopy $F \oplus G : f \simeq h$ *satisfying* $F \oplus G = F + G$ *on* $A_{(k+1)}$. *There is also* $\ominus F : g \simeq f$ *satisfying* $\ominus F = -F$ *on* $A_{(k+1)}$.

These constructions are natural in (A, V) *and* B.

The latter means $\beta(F \oplus G)\alpha = \beta F \alpha \oplus \beta G \alpha$ resp. $\beta(\ominus F)\alpha = \ominus \beta F \alpha$, for maps $\alpha : \tilde{A} \to A$ with $\alpha(\tilde{V}) \subset V$ and $\beta : B \to B'$.

Proof. Let $IA = \mathsf{T}(V^0 \oplus V^1 \oplus sV)$, D be the Baues-Lemaire cylinder (4.3). Factoring out the ideal generated by $v^0 - v^1$ and sv ($v \in V_{<k}$) yields the dgA $\bar{I}A = \mathsf{T}(V_{<k} \oplus V^0_{\geqslant k} \oplus V^1_{\geqslant k} \oplus (sV)_{>k})$, $\bar{D}$. Passing to the quotient, we obtain dgA-maps $\bar{\imath}_0, \bar{\imath}_1 : A \rightrightarrows \bar{I}A$ and $\bar{I}A \overset{\sim}{\leftarrow} A : \bar{q}$ (this is a cylinder object for A rel $A_{(k)}$, cf. [B]), and an $(\bar{\imath}_0, \bar{\imath}_1)$-derivation $\bar{S} : A \to \bar{I}A$. Let $\bar{I}A \amalg_A IA$ be the push-out of $\bar{\imath}_1 : A \to \bar{I}A$ along $i_0 : A \to IA$; it contains $\bar{I}A$ and IA as sub-dgAs. For the first part of the lemma it suffices to construct a natural lifting f for the diagram

$$\begin{array}{ccc} A \amalg A & \xrightarrow{(\bar{\imath}_0, i_1)} & \bar{I}A \amalg_A IA \\ {\scriptstyle (i_0, i_1)} \big\downarrow & \overset{\varphi}{\nearrow} & \big\downarrow {\scriptstyle \sim \ (\bar{q}, q)} \\ IA & \xrightarrow{q} & A \end{array}$$

such that $\Phi = \bar{S} + S$ on $V_{\leqslant k}$ (here $\Phi = \varphi S$ is the derivation homotopy accociated with φ). Then put $F \oplus G = (F, G)\,\Phi$; we have $F \oplus G = F + G$ on $V_{\leqslant k}$, hence on $A_{(k+1)}$ as desired. (Note that both $F \oplus G$ and $F + G$ are (f, f)-derivations on $A_{(k+1)}$.)

We construct the derivation homotopy $\Phi : A \to \bar{I}A \amalg_A IA$ from $\bar{\imath}_0$ to i_1 as follows. Put $\Phi = \bar{S} + S$ on $V_{\leqslant k}$ and extend this uniquely to an $(\bar{\imath}_0, i_1)$-derivation on $A_{(k+1)}$. Then

$$\begin{aligned} d\,\Phi(v) + \Phi\, d(v) &= (d\,\bar{s}v + d\,sv) + S\,dv \\ &= (\bar{\imath}_0 v - \bar{\imath}_1 v - \bar{S}\,dv) + (i_0 v - i_1 v - S\,dv) + S\,dv \\ &= \bar{\imath}_0(v) - i_1(v) \end{aligned}$$

for $v \in V_{\leqslant k}$, where we use $\bar{S} = 0$ on $A_{(k)}$. Hence Φ is a derivation homotopy $\bar{\imath}_0 \simeq i_1$ on $A_{(k+1)}$ satisfying $(\bar{q}, q)\,\Phi = 0$. There is an extension of Φ defined on A retaining these properties; a natural construction is as follows. Observe that $(\bar{q}, q)$ has the section i_1, and there is a natural derivation homotopy (thus chain homotopy) $h : i_1\,(\bar{q}, q) \simeq 1$, cf. (4.3). Define inductively $\Phi(v) = h(-\Phi\, d + \bar{\imath}_0 - i_1)(v)$ for $v \in V_{>k}$. Then Φ is clearly natural in the desired sense. — The construction of $\ominus F$ is similar. □

NOTATIONS

The following list contains notations used for certain objects, resp. for the categories evidently formed by these objects, resp. for certain subcategories.

AA	augmented algebra
LA	(non-graded) Lie algebra
gL	graded Lie algebra
DG	differential graded (over the integers)
dg	differential graded (positively = non-negatively)
dg*	differential graded (negatively = non-positively)
DGC	DG coalgebra
cDGC	cocommutative DG coalgebra
dgA	dg algebra
cdgA	commutative dg algebra
dgL	dg Lie algebra
dgH	(cocommutative) dg Hopf algebra
dgHh	(reduced) dg Hopf algebra up to homotopy
dgH′	(commutative) dg Hopf coalgebra
dg*A	dg* algebra
cdg*A	commutative dg* algebra
⋮	⋮
s	simplicial
sGp	simplicial group
sAlg	simplicial algebra
sAA	simplicial augmented algebra
sCoalg	simplicial coalgebra
sCC	simplicial coaugmented coalgebra
⋮	⋮
$F_.$	prolongation of functor F (to corresponding simplicial categories)
$\otimes$	prolongation of functor $\otimes$ (bifunctor of simplicial modules)

$\mathcal{S}$	category of spaces (= simplicial sets)
$\mathcal{C}^{\mathcal{D}}$	category of functors $\mathcal{D} \to \mathcal{C}$
${}^{\mathcal{D}}\mathcal{C}$	category of cofunctors $\mathcal{D} \to \mathcal{C}$
$\langle \mathrm{s} \rangle_{\mathcal{C}} = {}^{\Delta}\mathcal{C}$	category of simplicial objects over category $\mathcal{C}$
$\langle \dots \rangle$	category formed by objects named or abbreviated '...'
$\langle \mathrm{Set} \rangle$	category of sets
$\langle \mathrm{sGp} \rangle$	category of sGps
$\langle \mathrm{DG} \rangle$	category of chain complexes (over additive category $\mathcal{V}$)
$\mathcal{M}od_{\mathsf{K}}$	category of K-modules
⋮	⋮
$\mathcal{C}_0$	pointed category canonically associated with the category $\mathcal{C}$
$\mathcal{C}_r$	category of r-reduced objects in $\mathcal{C}_0$ (suppose here the objects in $\mathcal{C}$ are graded by integers, then $X \in \mathcal{C}_0$ is said to be *r-reduced* if $* \to X$ is an isomorphism in degrees $< r$; here $*$ is an initial-terminal object of $\mathcal{C}_0$)
$\mathcal{S}_0$	category of pointed spaces
$\langle \mathrm{dgA} \rangle_0$	category of augmented dgAs
$\langle \mathrm{dgC} \rangle_0$	category of coaugmented dgCs
⋮	⋮
$\mathcal{S}_r$	category of r-reduced spaces
$\mathcal{S}'_r$	... with rational homology of finite type
$\langle \mathrm{dgHh} \rangle_r$	category of r-reduced dgHhs
$\langle \mathrm{dgHh} \rangle'_r$	... with homology of finite type
⋮	⋮
$\langle \mathrm{cdg^*A} \rangle'_{\langle r \rangle}$	category of cdg*As with r-reduced cohomology of finite type
$\mathbf{dgL}_r$	category of r-reduced free dgLs
$\mathbf{dgL}'_r$	... with homology of finite type
$\mathbf{dgHh}_r$	category of r-reduced free dgHhs
$\mathbf{dgHh}'_r$	... with homology of finite type
$\mathbf{M}'_r$	category of r-reduced KS-algebras of finite type

Bibliography

[Ad] Adams J.F., *On the cobar construction*, Proc. Nat. Acad. Sci. **42** (1956), 409–412.

[Ad Hi] Adams J.F., Hilton P.J., *On the chain algebra of a loop space*, Comm. Math. Helv. **30** (1955), 305–330.

[A L] Aubry M., Lemaire J.–M., *Homotopies d'algèbre de Lie et de leurs algèbres enveloppantes*, in: Algebraic Topology—Rational Homotopy, L.N.M. 1318, Springer-Verlag (1988), 26–30.

[An] Anick D.J., *Hopf algebras up to homotopy*, J. of the AMS **2** (1989), 417–453.

[An 2] Anick D.J., *R-local homotopy theory*, L.N.M. 1418, Springer-Verlag (1988), 78–85.

[And] Anderson D.W., *Axiomatic homotopy theory*, in: Algebraic Topology, Waterloo 1978, L.N.M. 741, Springer-Verlag (1979), 520–547.

[B] Baues H.J., *Algebraic Homotopy*, Cambridge Stud. Adv. Math. 15, Cambridge Univ. Press (1989).

[B 2] Baues H.J., *Geometry of loop spaces and the cobar construction*, Memoirs of the AMS 230 (1980).

[Bki] Bourbaki N., *Lie groups and Lie algebras*, Part I, Hermann, Paris (1975).

[Bki 2] Bourbaki N., *Commutative algebra*, Hermann, Paris (1972).

[B L] Baues H.J., Lemaire J.M., *Minimal models in homotopy theory*, Math. Ann. **225** (1977), 219–242.

[Bou] Bousfield A.K., *The localization of spaces with respect to homology*, Topology **14** (1975), 133–150.

[Bou G] Bousfield A.K., Gugenheim V.K.A.M., *On PL De Rham theory and rational homotopy type*, Memoirs of the AMS 179 (1976).

[Bou K] Bousfield A.K., Kan D.M., *Homotopy limits, completions and localizations* L.N.M. 304, Springer Verlag (1972)

[Ca] Cartan H., *Théories cohomologiques*, Invent. Math. **35** (1976), 261–271.

[Ce P] Cenkl B., Porter R., *De Rham theorem with cubical forms*, Pac. J. Math. **112** (1984), 35–47.

[C M N] Cohen F.R., Moore J.C., Neisendorfer J., *Torsion in homotopy groups*, Ann. of Math. **109** (1979), 121–168.

[Cu] Curtis E.B., *Some relations between homotopy and homology*, Ann. of Math. **83**, (1965), 386–413.

[Cu 2] Curtis E.B., *Simplicial homotopy theory*, Adv. in Math. **6** (1971), 107–209

[Dw] Dwyer W.G., *Tame homotopy theory*, Topology **18** (1979), 321–338.

[E Mc] Eilenberg S., Mac Lane S., *On the groups* $H(\pi, n)$ (II), Ann. of Math. **60**, 1954, 49–139.

[E M] Eilenberg S., Moore J.C., *Limits and spectral sequences*, Topology **1** (1962), 1–24.

[Ep] Epstein D.B.A., *Semisimplicial objects and the Eilenberg-Zilber theorem*, Invent. math. **1** (1966), 209–220.

[Ep 2] Epstein D.B.A., *Functors between tensored categories*, Invent. math. **1** (1966), 221–228.

[FU] Boullay P., Kiefer F., Majewski M., Stelzer M., Scheerer H., Unsöld M., Vogt E., *Tame homotopy theory via differential forms*, Preprint n° 223 (1986), FU Berlin - FB Mathematik.

[FU 2] Scheerer H., Schuch K., Vogt E., *Tame homotopy theory via de Rham currents*, preprint n° A 91–20, FU Berlin (1991).

[G May] Gugenheim V.K.A.M., May J.P., *On the theory and application of differential torsion products*, Memoirs of the AMS 142 (1974).

[G Mu] Gugenheim V.K.A.M., Munkholm H.J., *On the extended functoriality of* Tor *and* Cotor, J. Pure and Appl. Algebra **4** (1974), 9–29.

[Gr Mg] Griffith P.A., Morgan J.W., *Rational homotopy theory and differential forms*, Progress in Math. 16, Birkhäuser (1981).

[Hal] Halperin S., *Lectures on minimal models*, Mémoires S.M.F. nouvelle série 9–10 (1983).

[He] Heller A., *Homotopy theories*, Memoirs of the AMS 383 (1988).

[Hess] Hess K., *Mild and tame homotopy theory*, J. Pure and Appl. Algebra **84** (1993), 277–310.

[Hi] Hilton P.J., *Note on quasi-Lie rings*, Fund. Math. **43** (1956), 230–237.

[H M S] Husemoller D., Moore J.C., Stasheff J., *Differential homological algebra and homogeneous spaces*, J. Pure and Appl. Algebra **5** (1974), 113–185.

[Hi St] Hilton P.J., Stammbach U., *A course in homological algebra*, GTM 4, Springer Verlag (1970).

[Jo] Jordan B.W., *A lower central series for split Hopf algebras with involution*, Trans AMS **257**, 427–454.

[K] Kan D.M., *On the homotopy relation for c.s.s. maps*, Bull. Soc. Math. Mexicana (1957), 75–81.

[K 2] Kan D.M., *On c.s.s. categories*, Bull. Soc. Math. Mexicana (1957), 82–94.

[K 3] Kan D.M., *A combinatorial definition of homotopy groups*, Ann. of Math. **67** (1958), 288–312.

[K 4] Kan D.M., *On homotopy theory and c.s.s. groups*, Ann. of Math. **68** (1958), 38–53.

[La S] Lambe L., Stasheff J., *Applications of perturbation theory to iterated fibrations*, manuscripta math. **58** (1987), 363–376.

[Laz] Lazard M., *Sur les groupes nilpotents et les anneaux de Lie*, Ann. Ec. Norm. Sup. **71** (1954), 101–190.

[Maj] Majewski M., *Tame homotopy theory via polynomial forms, part 1*, Diplomarbeit, FU Berlin (1988).

[Maj 2] Majewski M., *A proof of the Baues - Lemaire conjecture in rational homotopy theory*, in: Proc. of the Winter School Geometry and Physics in Srni (Jan. 1991), Rendiconti del Circolo Mat. di Palermo (serie II) **30** (1993), 113–123.

[May] May J.P., *Simplicial objects in algebraic topology*, Van Nostrand Math. Studies 11 (1967).

[Mc] Mac Lane S., *Categories for the working mathematician*, Springer Verlag, New York (1971).

[Mc 2] Mac Lane S., *Homology*, Grundlehren 114, Springer Verlag (1967).

[Mc 3] Mac Lane S., *Categorical algebra*, Bull. Amer. Math. Soc. **71** (1965), 40–106.

[Mi M] Milnor J., Moore J.C., *On the structure of Hopf algebras*, Ann. of Math. (2) **81** (1965), 211–264.

[Mo Pr] Morace F., Prouté A., *Brown's natural twisting cochain and the Eilenberg-Mac Lane transformation*, Prépublic. math. univers. Paris VII No. 29 (1992).

[Mu] Munkholm H., *DGA algebras as a Quillen model category, relations to shm maps*, J. Pure Appl. Algebra **13** (1978), 221–332.

[Pa] Passman D.S., *The algebraic structure of group rings*, Pure and applied Math., Wiley (1977).

[Pr] Prouté A., *A^∞-structures. Modèle de Baues - Lemaire et homologie des fibrations*, preprint.

[Q] Quillen D.G., *Rational homotopy theory*, Ann. of Math. (2) **90** (1969), 205–295.

[Q 2] Quillen D.G., *Homotopical algebra*, L.N.M. 43, Springer-Verlag (1967).

[Q 3] Quillen D.G., *An application of simplicial profinite groups*, Comment. Math. Helv. **44** (1969), 45–60.

[Q 4] Quillen D.G., *On the associated graded ring of a group ring*, J. algebra **10** (1968), 411–418.

[Se] Serre J.-P., *Lie algebras and Lie groups*, Benjamin, New York (1965).

[Sh T] Scheerer H., Tanré D., *The Milnor - Moore theorem in tame homotopy theory*, manuscr. math. **70** (1991), 227–246.

[Sh T 2] Scheerer H., Tanré D., *Exploring W.G. Dwyer's tame homotopy theory*, Publicacions Matemàtiques **35** (1991), 375–402.

[Sh T 3] Scheerer H., Tanré D., *R-local homotopy theory as part of tame homotopy theory*, Bull. London Math. Soc. **22** (1990), 591–598.

[Sh T 4] Scheerer H., Tanré D., *Homotopie modérée et tempérée avec les coalgèbres. Applications aux espaces fonctionnels*, Arch. Math. **59** (1992), 130–145.

[Shu] Schuch K., *Zahme Homotopietheorie mit cofiltrierten Lie Algebren und ein Milnor-Moore Theorem*, thesis, FU Berlin (1989).

[Ste] Stelzer M., *Tame homotopy theory via polynomial forms, part 2*, Diplomarbeit, FU Berlin (1989).

[Su] Sullivan D., *Infinitesimal computations in topology*, Inst. Haut. Etud. Sci. Publ. Math. **47** (1978), 269–331.

[T] Tanré D., *Homotopie rationnelle, Modèles de Chen, Quillen, Sullivan*, L.N.M. 1025, Springer Verlag (1983).

[Un] Unsöld H.M., *Differential torsion and fibrations*, Preprint, FU Berlin (1990).

[War] Warfield R.B. (Jr.), *Nilpotent groups*, L.N.M. 513, Springer Verlag (1976).

Editorial Information

To be published in the *Memoirs*, a paper must be correct, new, nontrivial, and significant. Further, it must be well written and of interest to a substantial number of mathematicians. Piecemeal results, such as an inconclusive step toward an unproved major theorem or a minor variation on a known result, are in general not acceptable for publication. *Transactions* Editors shall solicit and encourage publication of worthy papers. Papers appearing in *Memoirs* are generally longer than those appearing in *Transactions* with which it shares an editorial committee.

As of September 30, 1999, the backlog for this journal was approximately 5 volumes. This estimate is the result of dividing the number of manuscripts for this journal in the Providence office that have not yet gone to the printer on the above date by the average number of monographs per volume over the previous twelve months, reduced by the number of issues published in four months (the time necessary for preparing an issue for the printer). (There are 6 volumes per year, each containing at least 4 numbers.)

A Copyright Transfer Agreement is required before a paper will be published in this journal. By submitting a paper to this journal, authors certify that the manuscript has not been submitted to nor is it under consideration for publication by another journal, conference proceedings, or similar publication.

Information for Authors and Editors

Memoirs are printed by photo-offset from camera copy fully prepared by the author. This means that the finished book will look exactly like the copy submitted.

The paper must contain a *descriptive title* and an *abstract* that summarizes the article in language suitable for workers in the general field (algebra, analysis, etc.). The *descriptive title* should be short, but informative; useless or vague phrases such as "some remarks about" or "concerning" should be avoided. The *abstract* should be at least one complete sentence, and at most 300 words. Included with the footnotes to the paper, there should be the 1991 *Mathematics Subject Classification* representing the primary and secondary subjects of the article. This may be followed by a list of *key words and phrases* describing the subject matter of the article and taken from it. A list of the numbers may be found in the annual index of *Mathematical Reviews*, published with the December issue starting in 1990, as well as from the electronic service e-MATH [**telnet e-MATH.ams.org** (or **telnet 130.44.1.100**). Login and password are **e-math**]. For journal abbreviations used in bibliographies, see the list of serials in the latest *Mathematical Reviews* annual index. When the manuscript is submitted, authors should supply the editor with electronic addresses if available. These will be printed after the postal address at the end of each article.

Electronically prepared papers. The AMS encourages submission of electronically prepared papers in $\mathcal{A}_{\mathcal{M}}\mathcal{S}$-TeX or $\mathcal{A}_{\mathcal{M}}\mathcal{S}$-LaTeX. The Society has prepared author packages for each AMS publication. Author packages include instructions for preparing electronic papers, the *AMS Author Handbook*, samples, and a style file that generates the particular design specifications of that publication series for both $\mathcal{A}_{\mathcal{M}}\mathcal{S}$-TeX and $\mathcal{A}_{\mathcal{M}}\mathcal{S}$-LaTeX.

Authors with FTP access may retrieve an author package from the Society's Internet node `e-MATH.ams.org` (130.44.1.100). For those without FTP access, the author package can be obtained free of charge by sending e-mail to pub@ams.org (Internet) or from the Publication Division, American Mathematical Society, P.O. Box 6248, Providence, RI 02940-6248. When requesting an author package, please specify $\mathcal{A}\mathcal{M}\mathcal{S}$-TEX or $\mathcal{A}\mathcal{M}\mathcal{S}$-LATEX, Macintosh or IBM (3.5) format, and the publication in which your paper will appear. Please be sure to include your complete mailing address.

Submission of electronic files. At the time of submission, the source file(s) should be sent to the Providence office (this includes any TEX source file, any graphics files, and the DVI or PostScript file).

Before sending the source file, be sure you have proofread your paper carefully. The files you send must be the EXACT files used to generate the proof copy that was accepted for publication. For all publications, authors are required to send a printed copy of their paper, which exactly matches the copy approved for publication, along with any graphics that will appear in the paper.

TEX files may be submitted by email, FTP, or on diskette. The DVI file(s) and PostScript files should be submitted only by FTP or on diskette unless they are encoded properly to submit through e-mail. (DVI files are binary and PostScript files tend to be very large.)

Files sent by electronic mail should be addressed to the Internet address pub-submit@ams.org. The subject line of the message should include the publication code to identify it as a Memoir. TEX source files, DVI files, and PostScript files can be transferred over the Internet by FTP to the Internet node e-math.ams.org (130.44.1.100).

Electronic graphics. Figures may be submitted to the AMS in an electronic format. The AMS recommends that graphics created electronically be saved in Encapsulated PostScript (EPS) format. This includes graphics originated via a graphics application as well as scanned photographs or other computer-generated images.

If the graphics package used does not support EPS output, the graphics file should be saved in one of the standard graphics formats—such as TIFF, PICT, GIF, etc.—rather than in an application-dependent format. Graphics files submitted in an application-dependent format are not likely to be used. No matter what method was used to produce the graphic, it is necessary to provide a paper copy to the AMS.

Authors using graphics packages for the creation of electronic art should also avoid the use of any lines thinner than 0.5 points in width. Many graphics packages allow the user to specify a "hairline" for a very thin line. Hairlines often look acceptable when proofed on a typical laser printer. However, when produced on a high-resolution laser imagesetter, hairlines become nearly invisible and will be lost entirely in the final printing process.

Screens should be set to values between 15% and 85%. Screens which fall outside of this range are too light or too dark to print correctly.

Any inquiries concerning a paper that has been accepted for publication should be sent directly to the Editorial Department, American Mathematical Society, P. O. Box 6248, Providence, RI 02940-6248.

Editors

This journal is designed particularly for long research papers (and groups of cognate papers) in pure and applied mathematics. Papers intended for publication in the *Memoirs* should be addressed to one of the following editors:

Ordinary differential equations, partial differential equations, and applied mathematics to JOHN MALLET-PARET, Division of Applied Mathematics, Brown University, Providence, RI 02912-9000; electronic mail: jmp@cfm.brown.edu.

Harmonic analysis, representation theory, and Lie theory to ROBERT J. STANTON, Department of Mathematics, The Ohio State University, 231 West 18th Avenue, Columbus, OH 43210-1174; electronic mail: stanton@math.ohio-state.edu.

Ergodic theory and dynamical systems to ROBERT F. WILLIAMS, Department of Mathematics, University of Texas at Austin, Austin, TX 78712-1082; e-mail: bob@math.utexas.edu

Real and harmonic analysis and geometric partial differential equations to WILLIAM BECKNER, Department of Mathematics, University of Texas at Austin, Austin, TX 78712-1082; e-mail: beckner@math.utexas.edu.

Algebra to CHARLES CURTIS, Department of Mathematics, University of Oregon, Eugene, OR 97403-1222 e-mail: cwc@darkwing.uoregon.edu

Algebraic topology and cohomology of groups to STEWART PRIDDY, Department of Mathematics, Northwestern University, 2033 Sheridan Road, Evanston, IL 60208-2730; e-mail: s_priddy@math.nwu.edu.

Differential geometry and global analysis to CHUU-LIAN TERNG, Department of Mathematics, Northeastern University, Huntington Avenue, Boston, MA 02115-5096; e-mail: terng@neu.edu.

Probability and statistics to RODRIGO BAÑUELOS, Department of Mathematics, Purdue University, West Lafayette, IN 47907-1968; e-mail: banuelos@math.purdue.edu.

Combinatorics and Lie theory to PHILIP J. HANLON, Department of Mathematics, University of Michigan, Ann Arbor, MI 48109-1003; e-mail: hanlon@math.lsa.umich.edu.

Logic to THEODORE SLAMAN, Department of Mathematics, University of California at Berkeley, Berkeley, CA 94720-3840; e-mail: slaman@math.berkeley.edu.

Number theory and arithmetic algebraic geometry to ALICE SILVERBERG, c/o Mathematisches Institut, Universitaet Erlangen–Nuernberg, Bismarckstraße 1 1/2, 91054 Erlangen, Germany; e-mail: silver@math.ohio-state.edu.

Complex analysis and complex geometry to DANIEL M. BURNS, Department of Mathematics, University of Michigan, Ann Arbor, MI 48109-1003; e-mail: dburns@math.lsa.umich.edu.

Algebraic geometry and commutative algebra to LAWRENCE EIN, Department of Mathematics, University of Illinois, 851 S. Morgan (M/C 249), Chicago, IL 60607-7045; e-mail: ein@uic.edu.

Geometric topology, knot theory, hyperbolic geometry, and general topoogy to JOHN LUECKE, Department of Mathematics, University of Texas at Austin, Austin, TX 78712-1082; e-mail: luecke@math.utexas.edu.

Partial differential equations and applied mathematics to BARBARA LEE KEYFITZ, Department of Mathematics, University of Houston, 4800 Calhoun, Houston, TX 77204-3476; e-mail: keyfitz@uh.edu

Operator algebras and functional analysis to BRUCE E. BLACKADAR, Department of Mathematics, University of Nevada, Reno, NV 89557; e-mail: bruceb@math.unr.edu

All other communications to the editors should be addressed to the Managing Editor, PETER SHALEN, Department of Mathematics, University of Illinois, 851 S. Morgan (M/C 249), Chicago, IL 60607-7045; e-mail: shalen@math.uic.edu.

Selected Titles in This Series

(Continued from the front of this publication)

651 **Lindsay N. Childs, Cornelius Greither, David J. Moss, Jim Sauerberg, and Karl Zimmermann,** Hopf algebras, polynomial formal groups, and Raynaud orders, 1998

650 **Ian M. Musson and Michel Van den Bergh,** Invariants under Tori of rings of differential operators and related topics, 1998

649 **Bernd Stellmacher and Franz Georg Timmesfeld,** Rank 3 amalgams, 1998

648 **Raúl E. Curto and Lawrence A. Fialkow,** Flat extensions of positive moment matrices: Recursively generated relations, 1998

647 **Wenxian Shen and Yingfei Yi,** Almost automorphic and almost periodic dynamics in skew-product semiflows, 1998

646 **Russell Johnson and Mahesh Nerurkar,** Controllability, stabilization, and the regulator problem for random differential systems, 1998

645 **Peter W. Bates, Kening Lu, and Chongchun Zeng,** Existence and persistence of invariant manifolds for semiflows in Banach space, 1998

644 **Michael David Weiner,** Bosonic construction of vertex operator para-algebras from symplectic affine Kac-Moody algebras, 1998

643 **Józef Dodziuk and Jay Jorgenson,** Spectral asymptotics on degenerating hyperbolic 3-manifolds, 1998

642 **Chu Wenchang,** Basic almost-poised hypergeometric series, 1998

641 **W. Bulla, F. Gesztesy, H. Holden, and G. Teschl,** Algebro-geometric quasi-periodic finite-gap solutions of the Toda and Kac-van Moerbeke hierarchies, 1998

640 **Xingde Dai and David R. Larson,** Wandering vectors for unitary systems and orthogonal wavelets, 1998

639 **Joan C. Artés, Robert E. Kooij, and Jaume Llibre,** Structurally stable quadratic vector fields, 1998

638 **Gunnar Fløystad,** Higher initial ideals of homogeneous ideals, 1998

637 **Thomáš Gedeon,** Cyclic feedback systems, 1998

636 **Ching-Chau Yu,** Nonlinear eigenvalues and analytic-hypoellipticity, 1998

635 **Magdy Assem,** On stability and endoscopic transfer of unipotent orbital integrals on p-adic symplectic groups, 1998

634 **Darrin D. Frey,** Conjugacy of Alt_5 and $\mathrm{SL}(2,5)$ subgroups of $\mathrm{E}_8(\mathbb{C})$, 1998

633 **Dikran Dikranjan and Dmitri Shakhmatov,** Algebraic structure of pseudocompact groups, 1998

632 **Shouchuan Hu and Nikolaos S. Papageorgiou,** Time-dependent subdifferential evolution inclusions and optimal control, 1998

631 **Ronnie Lee, Steven H. Weintraub, and J. William Hoffman,** The Siegel modular variety of degree two and level four/Cohomology of the Siegel modular group of degree two and level four, 1998

630 **Florin Rădulescu,** The Γ-equivariant form of the Berezin quantization of the upper half plane, 1998

629 **Richard B. Sowers,** Short-time geometry of random heat kernels, 1998

628 **Christopher K. McCord, Kenneth R. Meyer, and Quidong Wang,** The integral manifolds of the three body problem, 1998

627 **Roland Speicher,** Combinatorial theory of the free product with amalgamation and operator-valued free probability theory, 1998

626 **Mikhail Borovoi,** Abelian Galois cohomology of reductive groups, 1998

For a complete list of titles in this series, visit the AMS Bookstore at **www.ams.org/bookstore/**.